Evolution of the Central Nervous System of Craniata and Homo

Wolfgang Seeger

Evolution of the Central Nervous System of Craniata and Homo

Wolfgang Seeger
Department of Neurosurgery
University Hospital Freiburg im Breisgau
Linden
Germany

ISBN 978-3-030-15218-5 ISBN 978-3-030-15216-1 (eBook)
https://doi.org/10.1007/978-3-030-15216-1

This Springer imprint is published by the registered company Springer Nature Switzerland AG
The registered company address is: Gewerbestrasse 11, 6330 Cham, Switzerland

Preface

A clearly arranged common basic training of studies of natural sciences was standard during more than 120 years after the examination of secondary schools. These were the base of a progredient interdisciplinary cooperation of the different natural sciences. Today, these teaching methods are penetrated by a medley of nonnatural scientific contents. Especially, subsidiary contents need cooperation with different other natural sciences, for example, neurosciences: the neurological main subject of all animal cells, the singular neuronal cell of animals, may present thousands of fine fiber structures. Each phylum presents its own similar neuronal cell-fiber system.

This book is divided into three parts:

Part I: Evolution of Craniata and Homo

The phylum of craniata presents a common medulla spinalis. Its medulla and its five divided encephalon of Pisces, Amphibia, Reptilia, Aves, and Mammalia were developed in more than 800 million years. It has adapted to the changing conditions of life during these long times. An exceptional position presents the duplicated telencephalon of Aves and further kinds of magnified allocortex since ca. 100 million years. The telencephalon of Mammalia presents a new type of telencephalon, the neocortex, combined with a reduced allocortex, since about 60–80 million years. The neocortex contains seven layers of pyramid and granular cells, variable in the different areas of the telencephalon.

Part II: Exceptional Position of the Human Encephalon

The most exceptional position presents the telencephalon of Mammalia, especially Homo sapiens. During its development from hominoides to Homo sapiens in only 2–3 million years, the weight of its telencephalon has been trebled. Its allocortical component is reduced to ca. less than 10%. It presents small areas of elementary special cortical centers, especially elementary psychological centers. These are taught for all children in schools and are the base for vocational training. Mathematics, philosophy, higher technologies, and other high psychological functions are located in the frontal telencephalon, but not exact to define. These functions are extremely different in different individuals. Large destructions of a frontal lobe (and preservation of the allocortex) may not be followed by psychological deficits. All parts of the telencephalon are combined with the allocortex. The diencephalon, mesencephalon, and rhombencephalon (part of the metencephalon) are systematic and bilateral symmetric structured with rare roundabout and interrupted fiber connections of all craniata. These are old portions in all craniatas. The cerebellum of craniata (part of the metencephalon of all craniata) is more systematically organized.

Part III: Morphologic Maturity and Immaturity

A high grade of maturity of all parts of the telencephalon presents Pisces (since ca. 800 million years), Amphibia (since ca. 400 million years), and Reptilia (since ca. 300 million years). The telencephalon of Aves is a new kind of telencephalon. Its rhinencephalon is smaller. Its other allocortical components are much wider than all other craniata in contrary to the old parts of the encephalon. It presents more roundabout fibers and less vulnerable than older parts of the brain. Only Mammalia are craniata, which presents a progredient other new kind of telencephalon, the neocortex. The progredient size of the neocortex is combined with the reduction of its allocortex. The neocortex of Homo presents more than 90% of the telencephalon, the allocortex less than 10%. The main fibers of neocortical neurons present the subcortical U-fibers with numerous double, treble, and roundabout fiber systems. Nearly all allocortical cell and fiber systems are systematically structured with rare duplicated or roundabout fibers. The neocortex presents extreme much roundabout fibers and fiber interruptions. *Morphologic human telencephalon is the most immature brain structure in the craniata. This is the beginning of the development of a new brain structure. It presents since historical times (10,000 years) similar extreme differences of psychological functions of its individuals. But most homines do not want to realize this fact.*

Linden, Germany
September 23th, 2018

Wolfgang Seeger

Contents

Part I Evolution of Craniata and Homo

1 Chronological Survey 3

2 Ontogenetic and Phylogenetic Basis 5
2.1 Survey 5
2.2 Details 9

3 Comparative Morphology of the Adult Central Nervous System of Craniata 17
3.1 Pisces 17
3.2 Amphibia 19
3.3 Reptilia and Aves (Both Together: Sauropsides) 20
3.4 Mammalia 21

Part II Exceptional Position of the Human Encephalon

4 Telencephalon, Survey 29

5 Diencephalon, Mesencephalon, Rhombencephalon, Cerebellum: Survey 35

6 Telencephalon: Details 39
6.1 Definition of Telencephalic Fiber Systems and Gyri: Introduction 39
6.1.1 Gyrification and U-Fibers 40
6.1.2 Association Fibers: Anatomical Fiber Dissections 45
6.1.3 Projection Fibers 48

7 Diencephalon, Mesencephalon, and Rhombencephalon: Details 57
7.1 Topography 57
7.2 Fiber Connections 61

8 Transectional Planes of Rhombencephalon (Medulla Oblongata, Pons) and Mesencephalon 65
8.1 Topography 65
8.2 Fiber Connections 72

9 Cranial Nerves 77
9.1 Principles of the Embryonic Developments: Survey 77
9.2 Cranial Nerves: Details 78

Part III Morphologic Maturity and Immaturity

10 Recent Neocortex 91
10.1 Neocortical Areas 91
10.2 Paleoanthropological Aspects 94

Bibliography 99

Part I

Evolution of Craniata and Homo

1 Chronological Survey

- 1.1 Pisces (since kambrium, silur, ca. 800–600 million years)
- 1.2 Amphibia (since perm, 350 million years)
- 1.3 Reptiles (since trias, jura, 250 million years)
- 1.4 Aves (cretaceus period, since 180 million years)
- 1.5 Mammalia (since tertiary, 70 million years)
  - 1.5.1 Monotremata (ovipara, ornitorhyncus)
  - 1.5.2 Marsupialia (vivipara, e.g., kangaroo)
  - 1.5.3 Placentalia since 60 million years
    - 1.5.3.1 Insectivora (shrew mouse, Erynaceus, mole, chiroptera)
    - 1.5.3.2 **Apatidae** since ca. 80 million years
      - 1.5.3.2.1 Prosimiae (e.g., *Tupaja glis*) since ca. 60 million years
      - 1.5.3.2.2 Simiae
      - 1.5.3.2.3 Hominoides since ca. 22 million years
        - 1.5.3.2.3.1 Orang utan since ca. 16 million years
        - 1.5.3.2.3.2 Gorilla since ca. 7 million years
        - 1.5.3.2.3.3 Chimpanzee since ca. 6 million years
        - 1.5.3.2.3.4 Homo since 2.5–2 million years
          - 1.5.3.2.3.4.1 *Homo rudolfensis* ("Lucy")
          - 1.5.3.2.3.4.2 *Homo habilis*
          - 1.5.3.2.3.4.3 *Homo ergaster*
          - 1.5.3.2.3.4.4 *Homo erectus* and others
            - 1.5.3.2.3.4.4.1 *Homo heidelbergensis*
            - 1.5.3.2.3.4.4.2 *Homo sapiens* (one common species)
              - 1.5.3.2.3.4.4.2.1 *Homo steinheimensis* ca. 350,000 years
              - 1.5.3.2.3.4.4.2.2 *Homo neanderthalensis* ca. 200,000–30,000 years
              - 1.5.3.2.3.4.4.2.3 *Homo recens* (Diluvium-Alluvium) since ca. 30,000 years
    - 1.5.3.3 Rodentia
    - 1.5.3.4 Carnivora
    - 1.5.3.5 Ungulata
    - 1.5.3.6 Cetacea (whales)

W. Seeger, *Evolution of the Central Nervous System of Craniata and Homo*, https://doi.org/10.1007/978-3-030-15216-1_1

2 Ontogenetic and Phylogenetic Basis

2.1 Survey (Figs. 2.1, 2.2, 2.3, and 2.4)

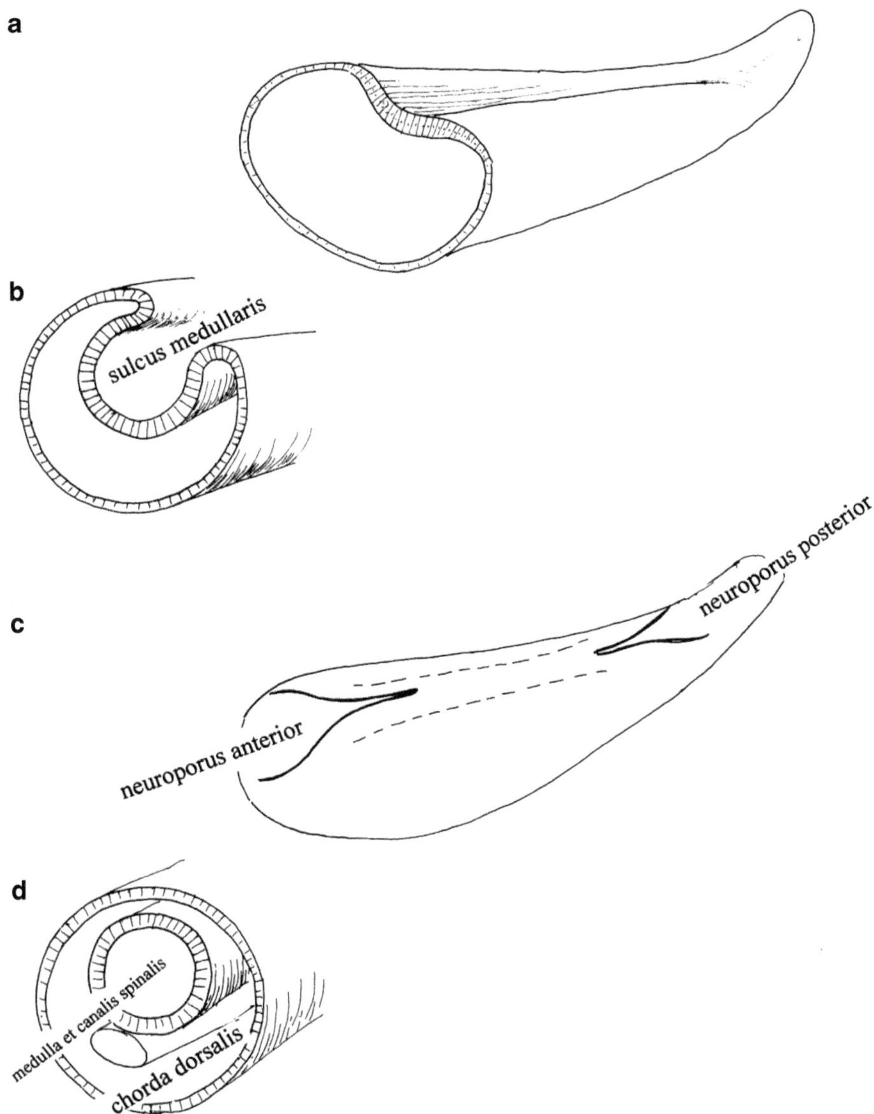

Fig. 2.1 *Acraniata, basis*: *First stage of ontogenesis and phylogenesis: Amphioxus*. (**a**) and (**b**) development of sulcus medullaris, (**c**) development of canalis medullaris (Dysraphic defects of human embryos originate during this step of development), (**d**) skeleton: chorda dorsalis. Tube of connective tissue is filled with mucosa (Residual in mammals is nucleus pulposus in intervertebral discs. Chordomas of the spinocranial systems in man develops from residuals of chorda dorsalis)

W. Seeger, *Evolution of the Central Nervous System of Craniata and Homo*, https://doi.org/10.1007/978-3-030-15216-1_2

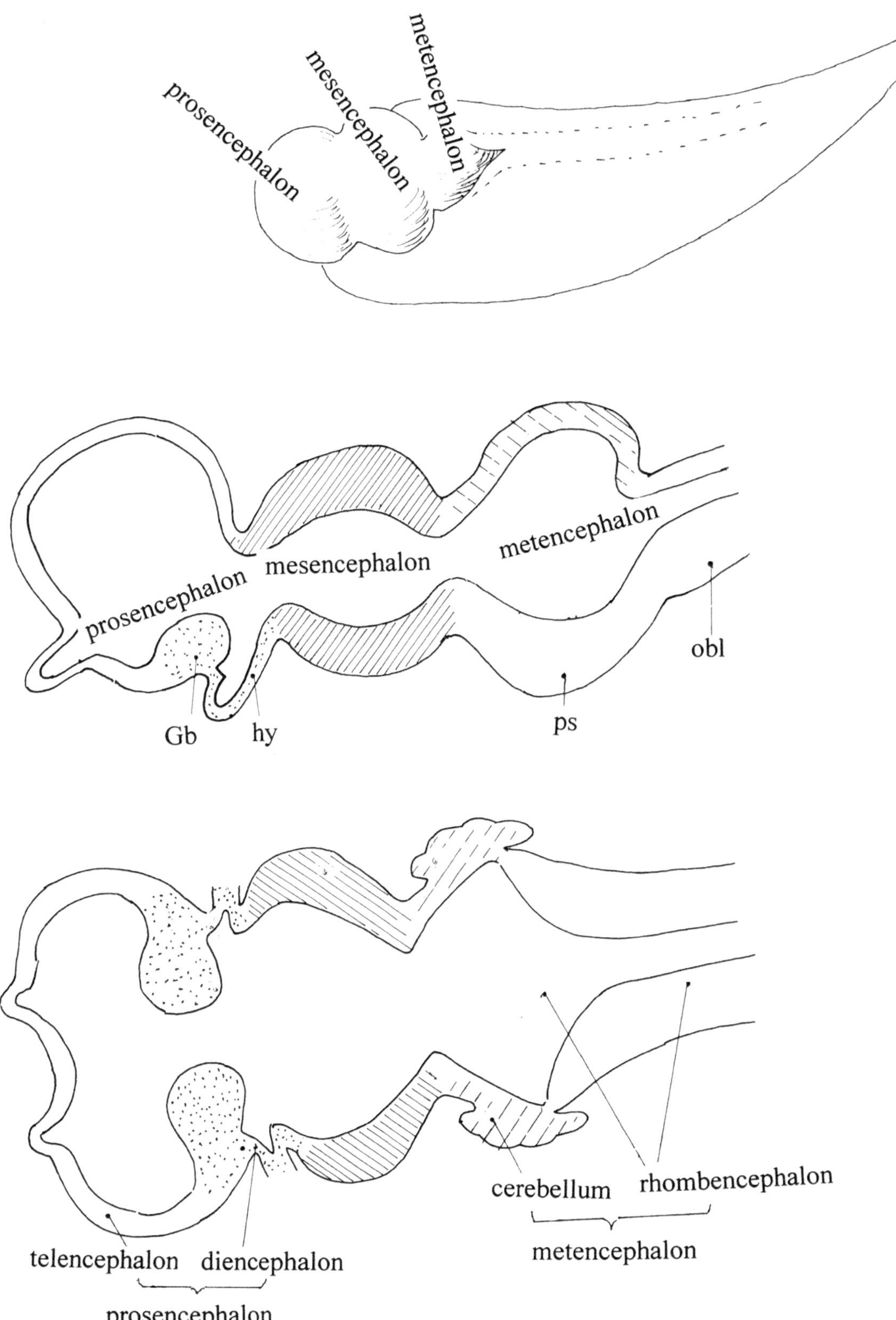

Fig. 2.2 *Craniata, basis*: First stage of ontogenesis and phylogenesis: At neuroporus anterior it develops a thin-walled roof, presenting three vesicles: prosencephalon, mesencephalon, and metencephalon. After this, the vesicles are widened to five vesicles. These five vesicles are the basis of the evolution of five compartments of encephalon in all craniata: (*1*) and (*2*) telencephalon duplicated, lateral ventricles; (*3*) diencephalon, third ventricle (*1–3* prosencephalon); (*4*) mesencephalon aquaeductus Sylvii; and (*5*) rhombencephalon and cerebellum, fourth ventricle (metencephalon). *Gb* gangliae basales, *hy* neurohypophysis, *ps* pons, *obl* medulla oblongata

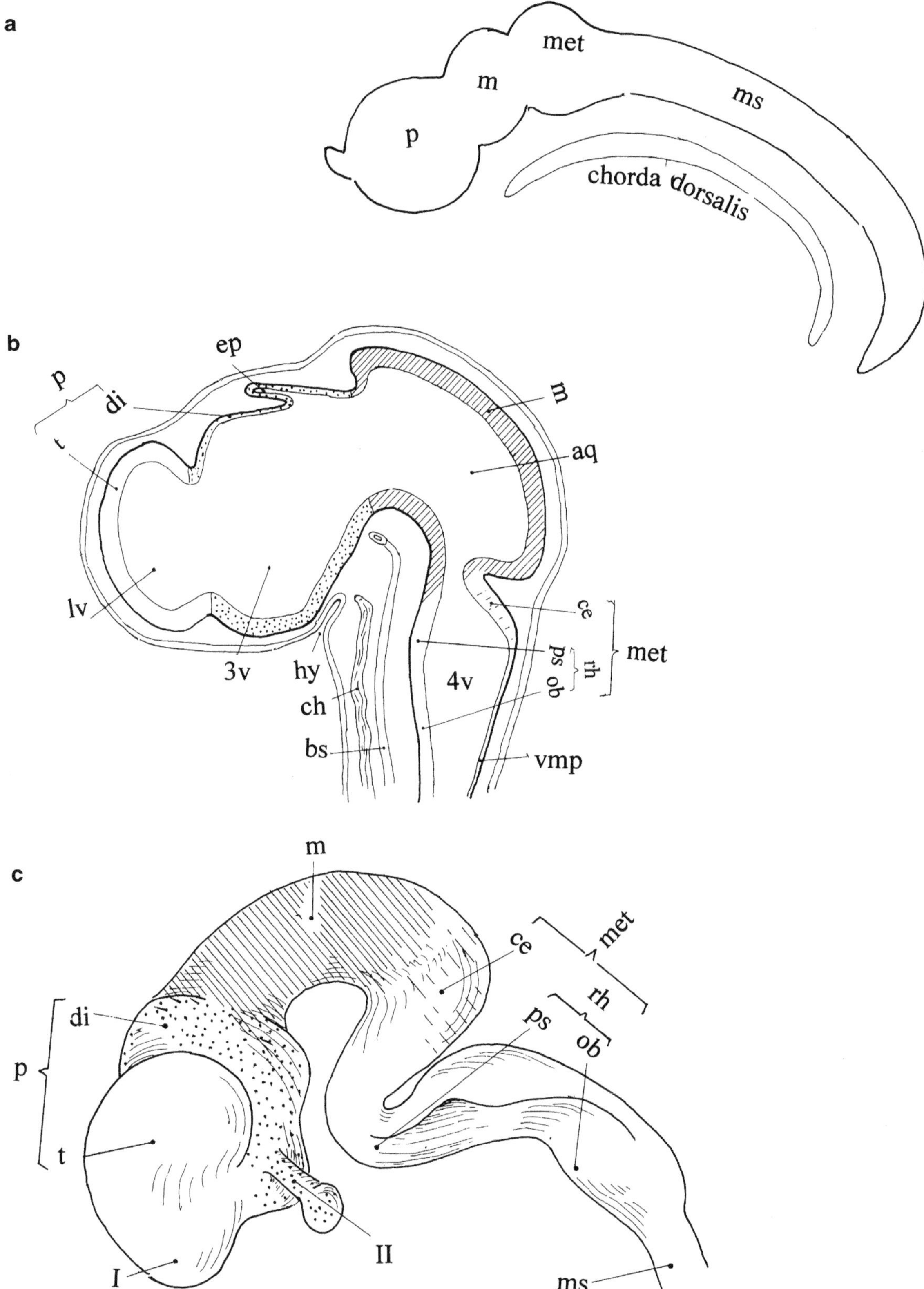

Fig. 2.3 *Examples*: **a**: three-ventricular stage; **b**: later. Five compartments of a chicken embryonic brain (V. v. Mihalkovics, cit. Rauber-Kopsch [7], vol. 5, p. 528, Indian ink copy, modified); **c**: seven weeks old *human embryo* (Rauber-Kopsch [7], vol. 5, p. 528, modified). *p* prosencephalon, *m* mesencephalon, *met* metencephalon, *ms* medulla spinalis, *t* telencephalon, *di* diencephalon, *ep* epiphysis, *aq* aquaeductus, *ce* cerebellum, *ps* pons, *rh* rhombencephalon, *vmp* velum medullare posterius, *4v* 4th ventricle, *ob* medulla oblongata, *bs a.* basilaris, *ch* chorda dorsalis, *hy* Rathke's excavation, *3v* third ventricle, *lv* lateral ventricle, (*I*) bulbus olfactorius, (*II*) fasciculus opticus

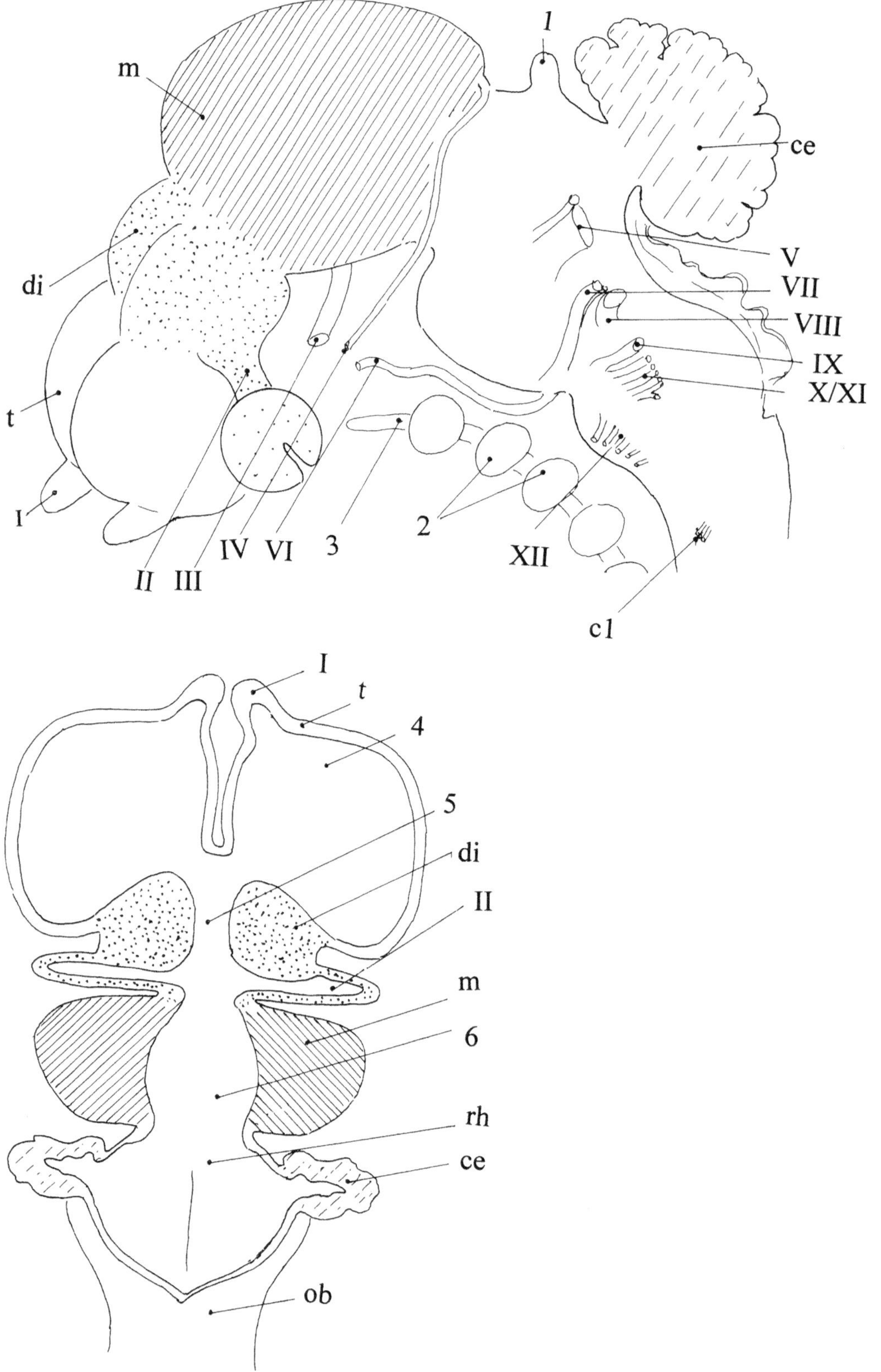

Fig. 2.4 *General embryonic architecture of encephalon in all terrestric craniata, Schematized*: *t* telencephalon, *di* diencephalon, *m* mesencephalon, *ce* cerebellum, *rh* rhombencephalon and fourth ventricle, *ps* pons, *ob* medulla oblongata, (I) bulbus olfactorius, (II) n. opticus, (III) n. oculomotorius, (IV) n. trochlearis, (V) n. trigeminus, (VI) n. abducens, (VII) n. facialis and intermedius, (VIII) n. statoacusticus, (IX) n. glossopharyngeus, (X) and (XI) n. vagus and n. accessorius, (XII) n. hypoglossus, (C1) n. spinalis 1, (*1*) corpus pineale, (*2*) somites, (*3*) chorda dorsalis, (*4*) ventriculus lateralis, (*5*) third ventricle, (*6*) aquaeductus

2.2 **Details** (Figs. 2.5, 2.6, 2.7, 2.8, 2.9, 2.10, 2.11, and 2.12)

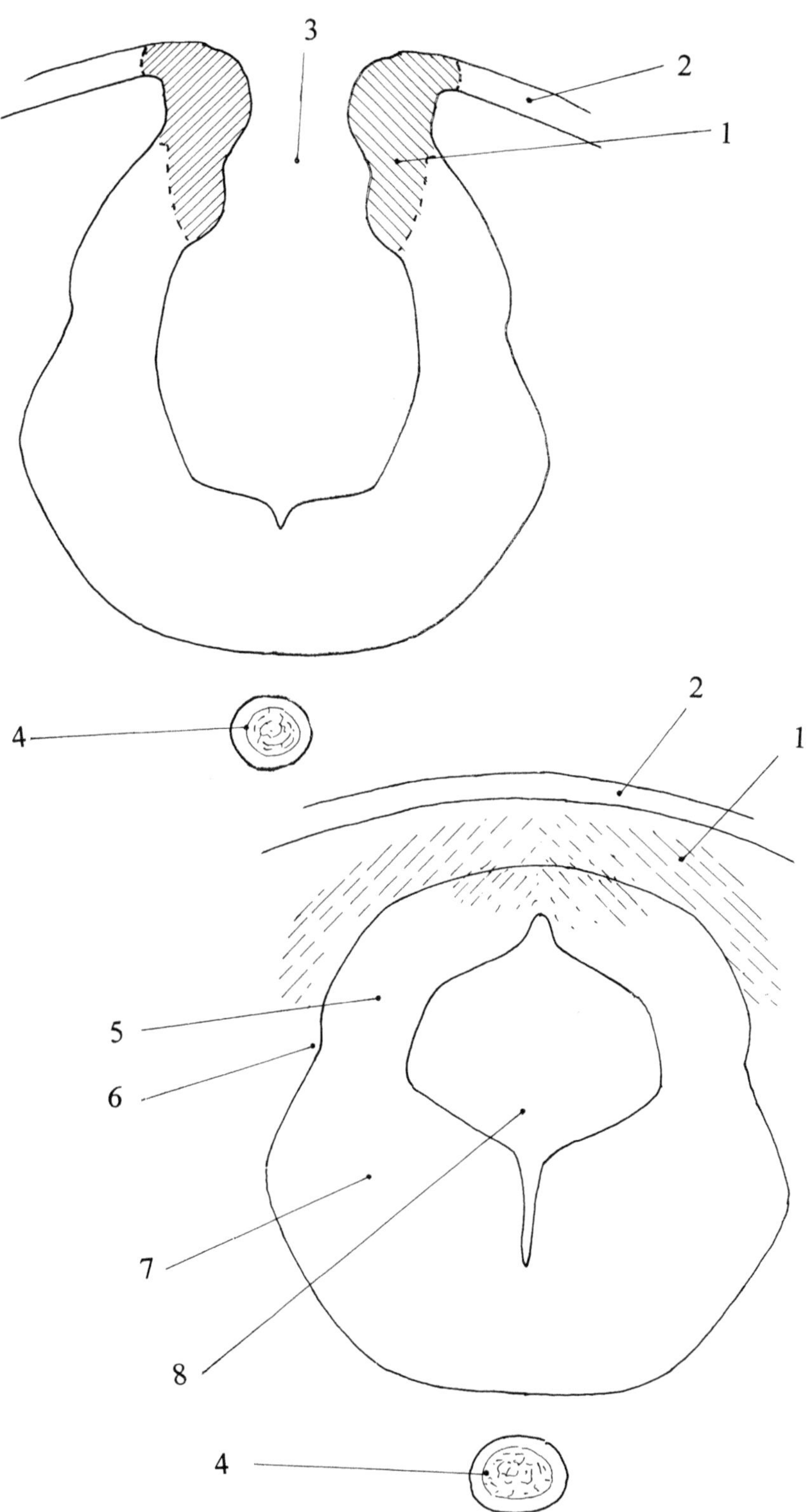

Fig. 2.5 *Medulla spinalis and adjacent structures. Early embryonic development*: Before and after closing of Sulcus medullaris into Canalis centralis, the bilateral neural crest is growing out from the neural fold. (*1*) Neural crest at neural fold, (*2*) ectoderm, (*3*) neural groove, (*4*) chorda dorsalis, (*5*) ala medullaris, (*6*) sulcus limitans, (*7*) basis medullaris, (*8*) canalis centralis

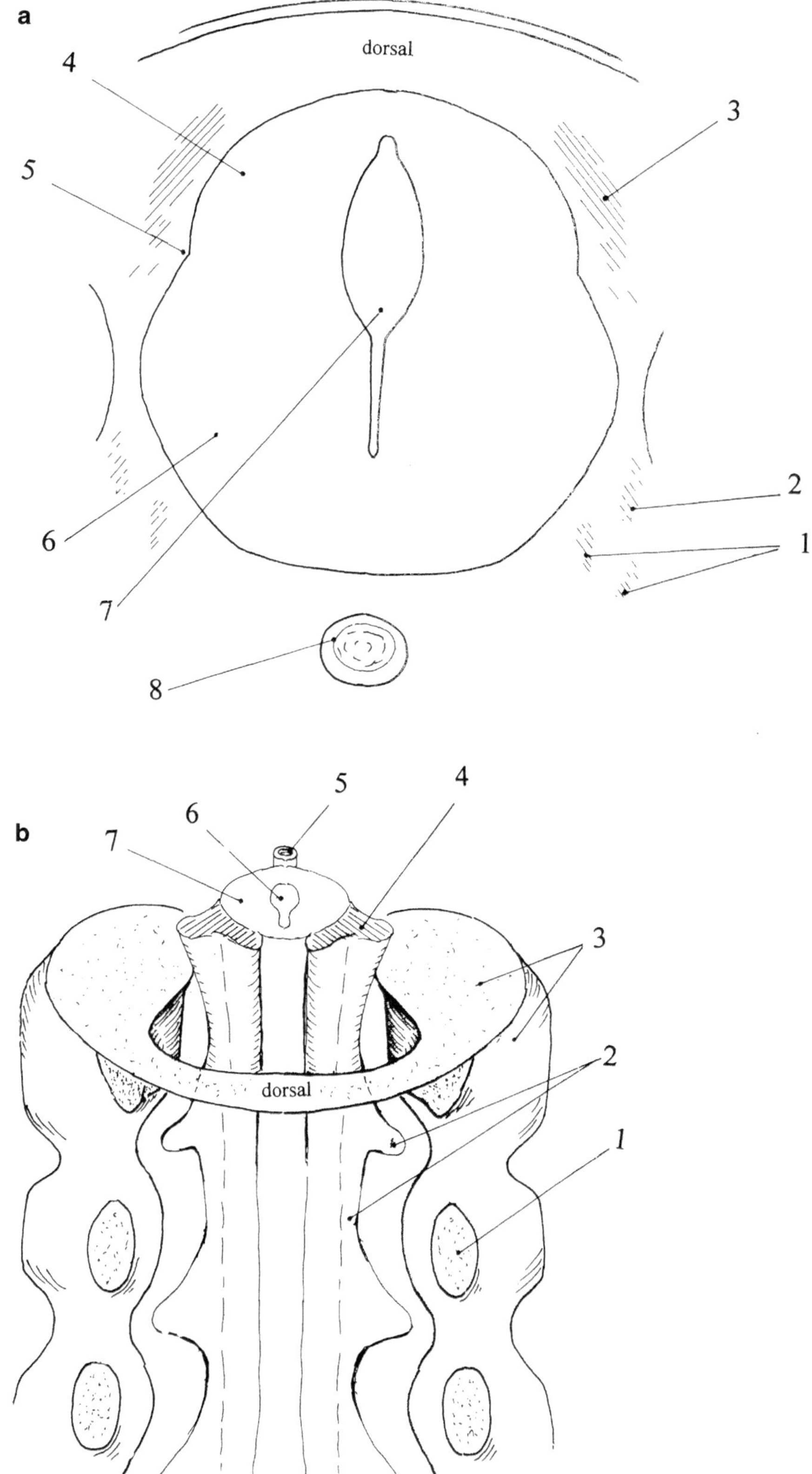

Fig. 2.6 *Continuation. Neural crests become loosening from medulla during the subdividing of somites. They are increasing between the somites and performing spinal ganglia, nerve roots, and spinal nerves.* **a**: (*1*) early ganglia mesenteriales, mixtum of parts of neurolines and mesodermal cells, (*2*) early gangliae paravertebrales, (*3*) mixtum of neuromeres and mesodermal cells: gangliae spinales, (*4*) ala medullaris, (*5*) sulcus limitans, (*6*) basis medullaris, (*7*) canalis centralis, (*8*) chorda dorsalis. **b**: after the development of somites and its transformation into vertebrae: (*1*) arcus vertebrae, cut, (*2*) dorsal neuroline, (*3*) somite, (*4*) as (*2*), (*5*) chorda dorsalis, (*6*) canalis centralis, (*7*) medulla spinalis

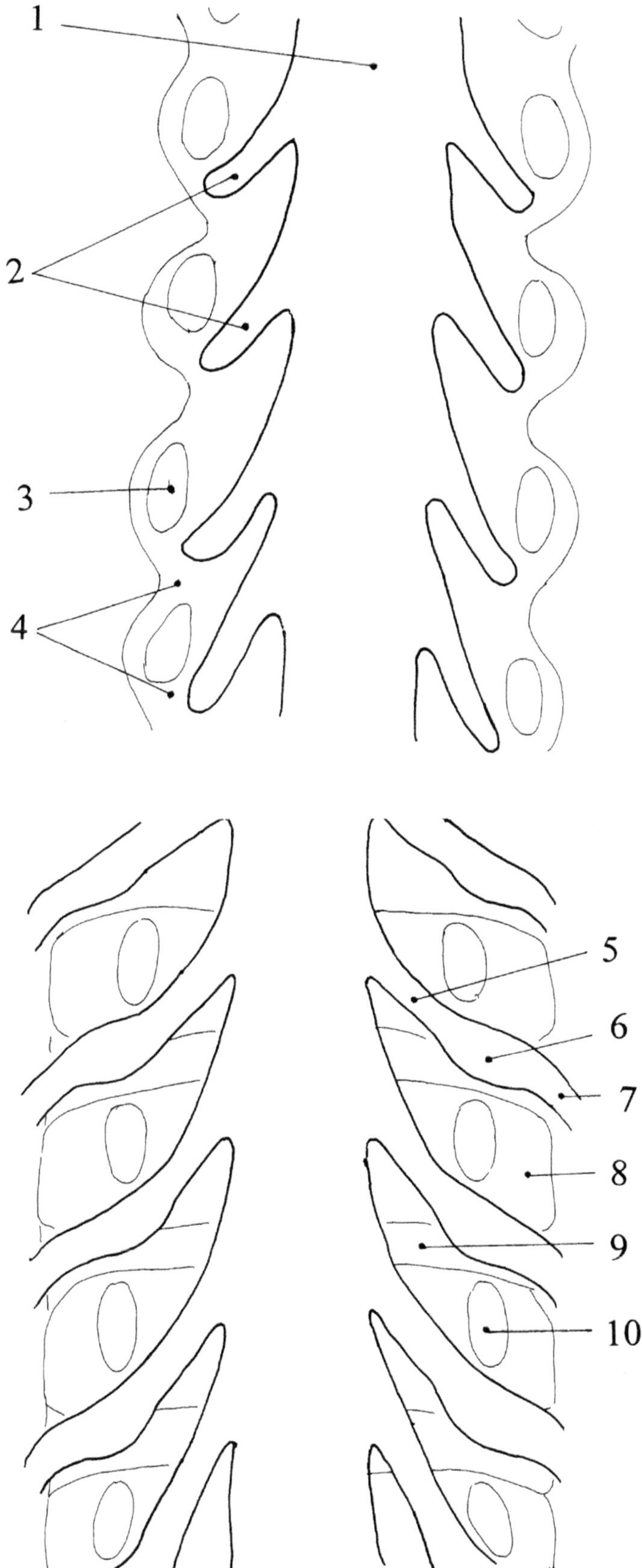

Fig. 2.7 *Addendum*. (*1*) Medulla spinalis, (*2*) neuromeres, (*3*) radix vertebrae cut, (*4*) somites/corpora vertebralia, (*5*) radix spinalis, (*6*) ganglion spinale, (*7*) n spinalis 1, (*8*) somite/corpus vertebrae, (*9*) discus intervertebralis, (*10*) radix vertebrae

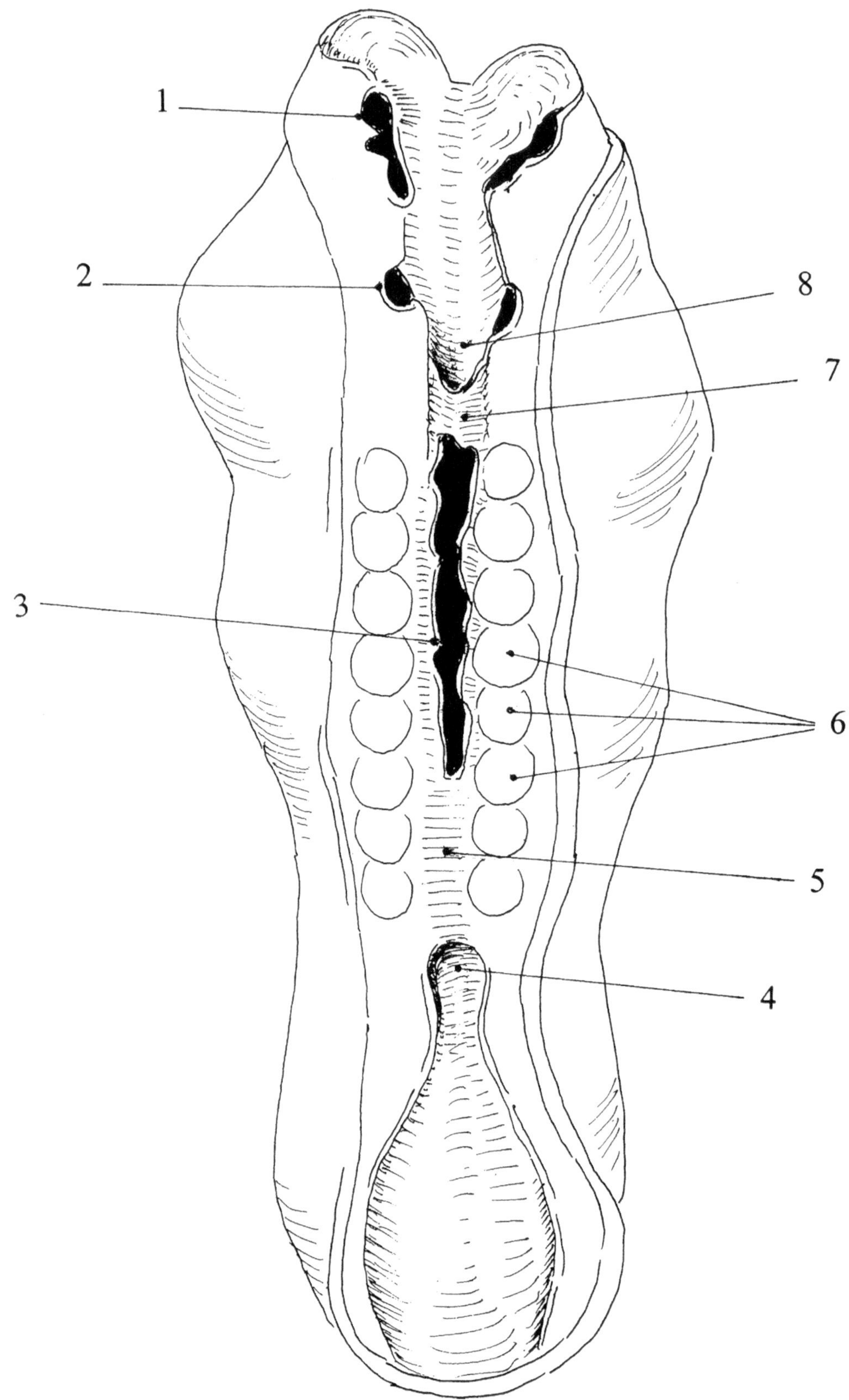

Fig. 2.8 *Spinocranial system* (Veit (1922), cit. Rauber/Kopsch [8], p. 26, Indian ink copy). *Development of somites in craniata (human embryo)*. Neuroporus anterior is not yet closed. (*1*) Rostral cranial neural crest (black colored) at neuronal fold, (*2*) caudal cranial neural crest (black colored), (*3*) dorsal neural crest (black colored), (*4*) neuroporus posterior, (*5*) roof of canalis medullaris, (*6*) 3 of 8 somites (Figs. 2.8 and 2.9), (*7*) as (*5*), (*8*) neuroporus anterior

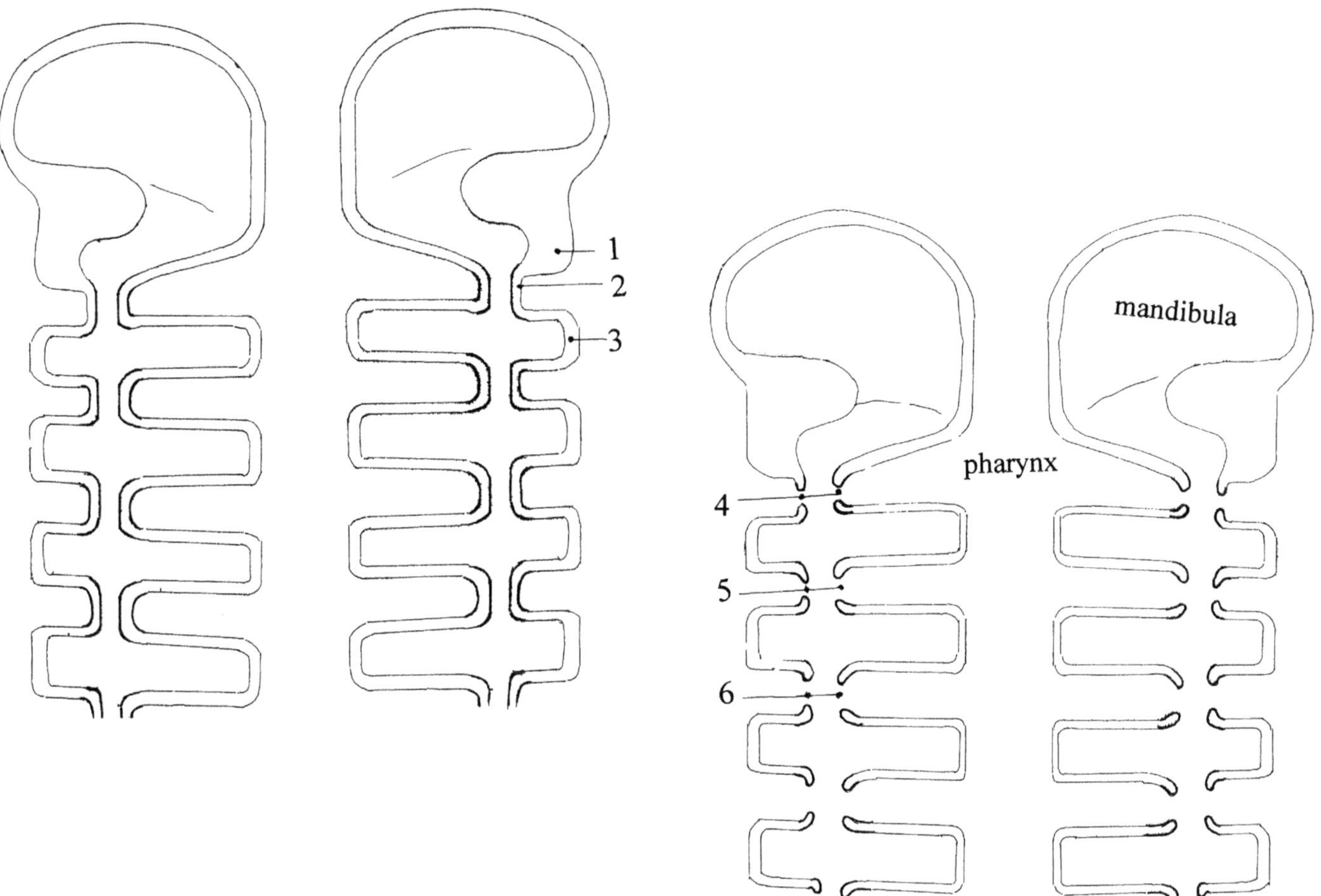

Fig. 2.9 *Addendum for spinocranial systems. Early development of branchial arches (gills) in pisces*. This is the basis of the development of craniopharyngeal structures in terrestric craniata. Nerves of branchial arches are the basis of caudal cranial nerves of terrestric craniata (except N. hypoglossus which is present in all craniata). (*1*) Processus mandibularis (first branchial arch according to N. trigeminus and its target areas), (*2*) ectodermal first branchial sulcus, (*3*) ectodermal second branchial arch (according to N. facialis and its target areas), (*4*) entodermal first branchial sulci, (*5*) entodermal second branchial sulci, (*6*) entodermal third branchial sulci. Nerves of the ectodermal third branchial arch are the basis of nn. vagus, glossopharyngeus, and accessorius of terrestric craniata

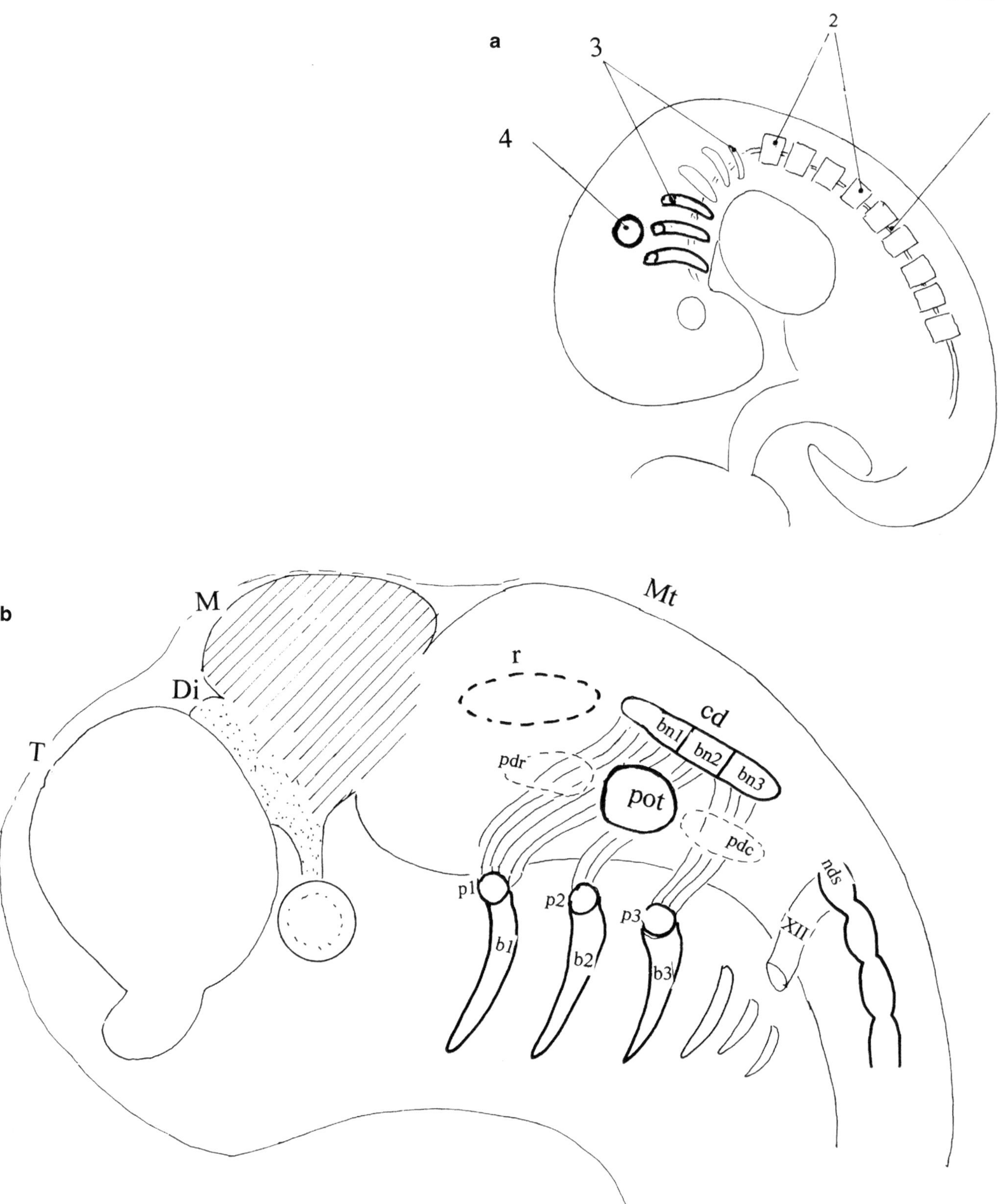

Fig. 2.10 *Transformation of branchial arches, neuromeres, and ectodermal plates of pisces into caudal cranial nerves and its target areas of terrestric craniata.* The early stages of the development of encephalon of pisces are present in the early embryonic stages of terrestric craniata. **a**: Embryo of pisces and terrestric craniata. (*1*) Chorda dorsalis, (*2*) somites, (*3*) branchial arches, (*4*) auricular patch of ectoderm (in pisces: lateral organ). **b**: Early stage of embryonic encephalon of pisces and terrestric craniata. *T* telencephalon, *Di* diencephalon, *M* mesencephalon, *Mt* metencephalon, *r* rostral cranial neural crests in pisces, *cd (bn1-bn3)* caudal cranial neural crests in all craniatas, *p1 to p3* neuronal crests, connected with ectodermal plates (pdr to pdc) and with branchial arches (b1 to b3), *pot* auriculovestibular patch, *XII* n. hypoglossus, *nds* neuronal crests for medulla spinalis

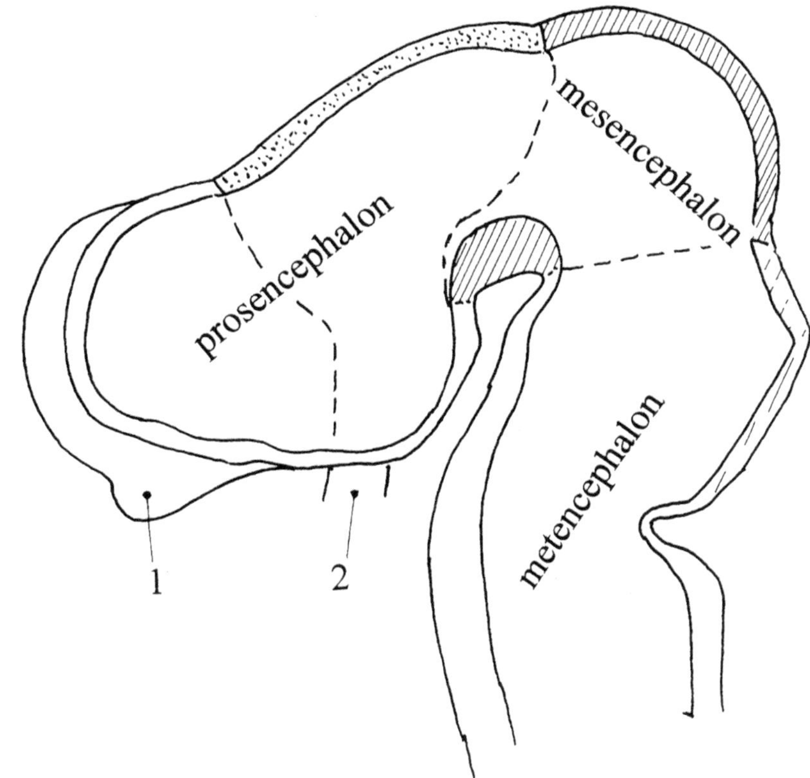

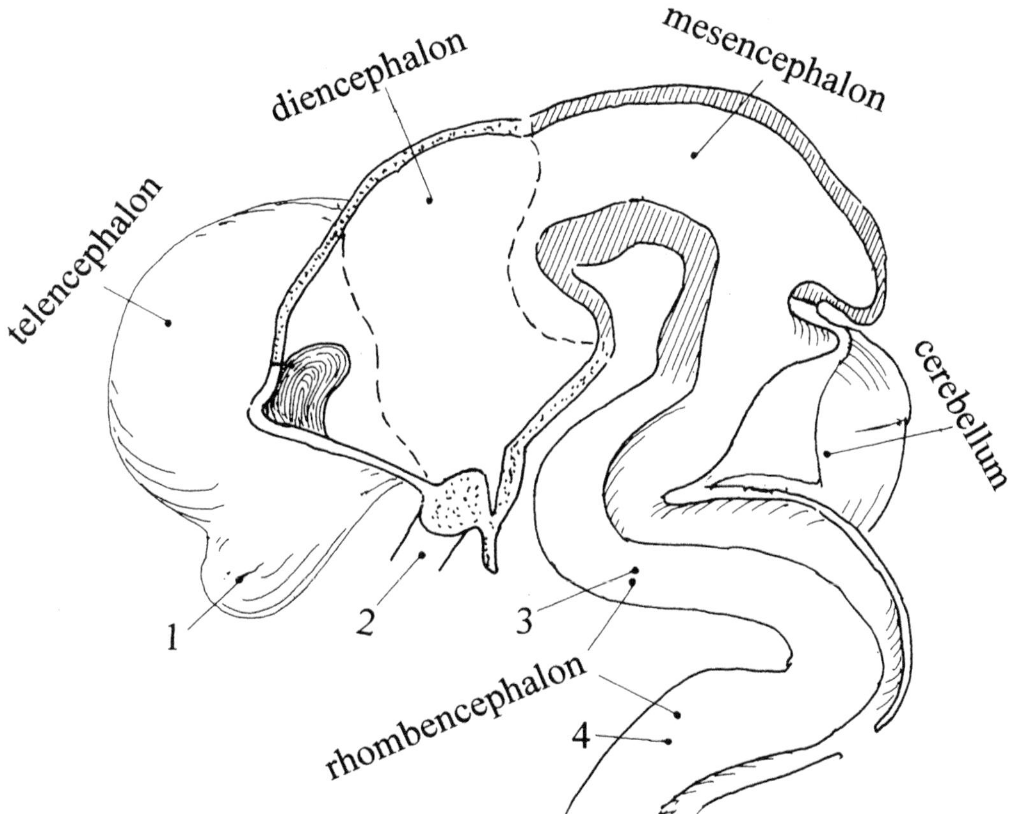

Fig. 2.11 *Embryonic development of Encephalon in mammalia (and homo)*, early stages, schematized, according to His (Cit. Spalteholz [12], vol. 3, p. 624). (*1*) Bulbus olfactorius, (*2*) N. opticus, (*3*) pons, (*4*) medulla oblongata

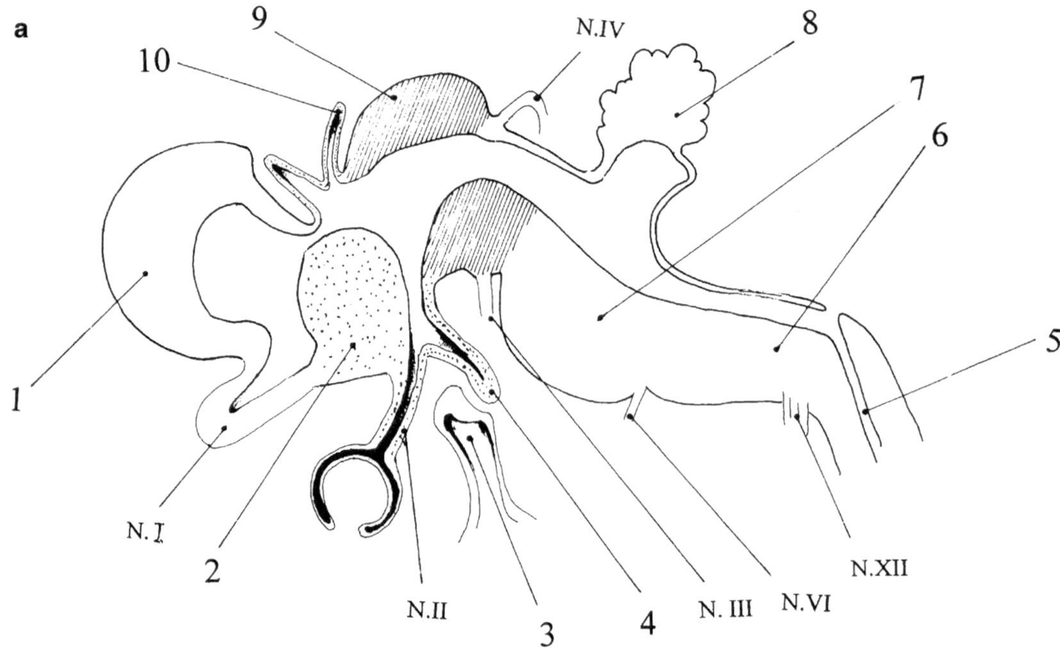

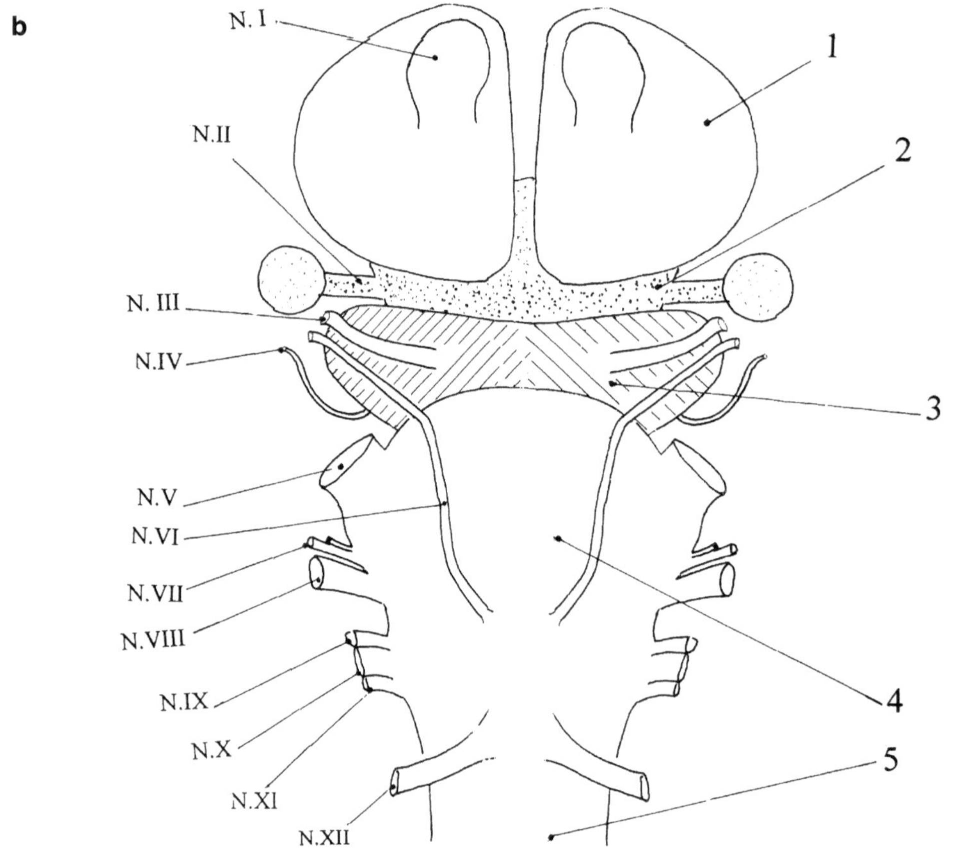

Fig. 2.12 *Later embryonic stage in mammalia, principles*. **a**: (*1*) telencephalon, (*2*) diencephalon (gangliae basales), (*3*) pharyngeal excavation (Rathke) according to adenohypophysis, (*4*) excavation of hypothalamus according to infundibulum and neurohypophysis, (*5*) canalis spinalis centralis, (*6*) medulla oblongata, (*7*) pons, (*8*) cerebellum, (*9*) mesencephalon, (*10*) epiphysis. **b**: (*1*) telencephalon, (*2*) diencephalon, (*3*) mesencephalon, (*4*) pons, (*5*) medulla oblongata

3 Comparative Morphology of the Adult Central Nervous System of Craniata

3.1 Pisces (Figs. 3.1 and 3.2)

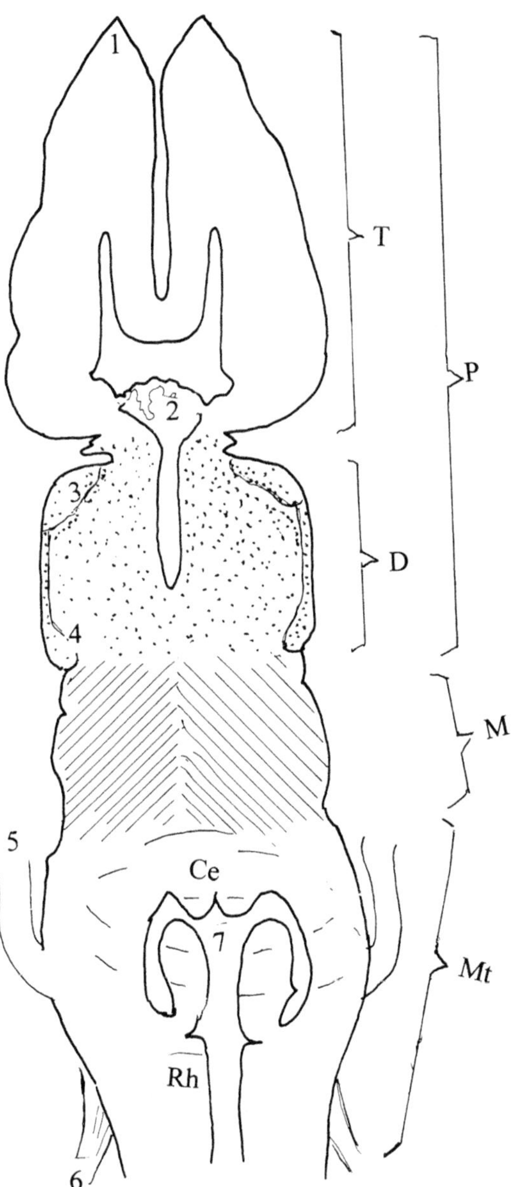

Fig. 3.1 *Selachian (shark) brain* (L. Edinger, cit. Rauber-Kopsch [7], vol. 5, p. 607. Modified Indian ink copy). Archaic type of recent pisces. Principles of five compartments of encephalon in all craniata. Axial transectional plane. *P* double prosencephalon, *T* telencephalon, *D* diencephalon, *M* mesencephalon, *Mt* metencephalon, *Ce* cerebellum, *Rh* rhombencephalon, (*1*) bulbus olfactorius, (*2*) Plexus chorioideus, (*3*) Tractus opticus, (*4*) Corpus geniculatum laterale, (*5*) nerve of the first branchial arch, (*6*) N. hypoglossus, (*7*) fourth ventricle

W. Seeger, *Evolution of the Central Nervous System of Craniata and Homo*, https://doi.org/10.1007/978-3-030-15216-1_3

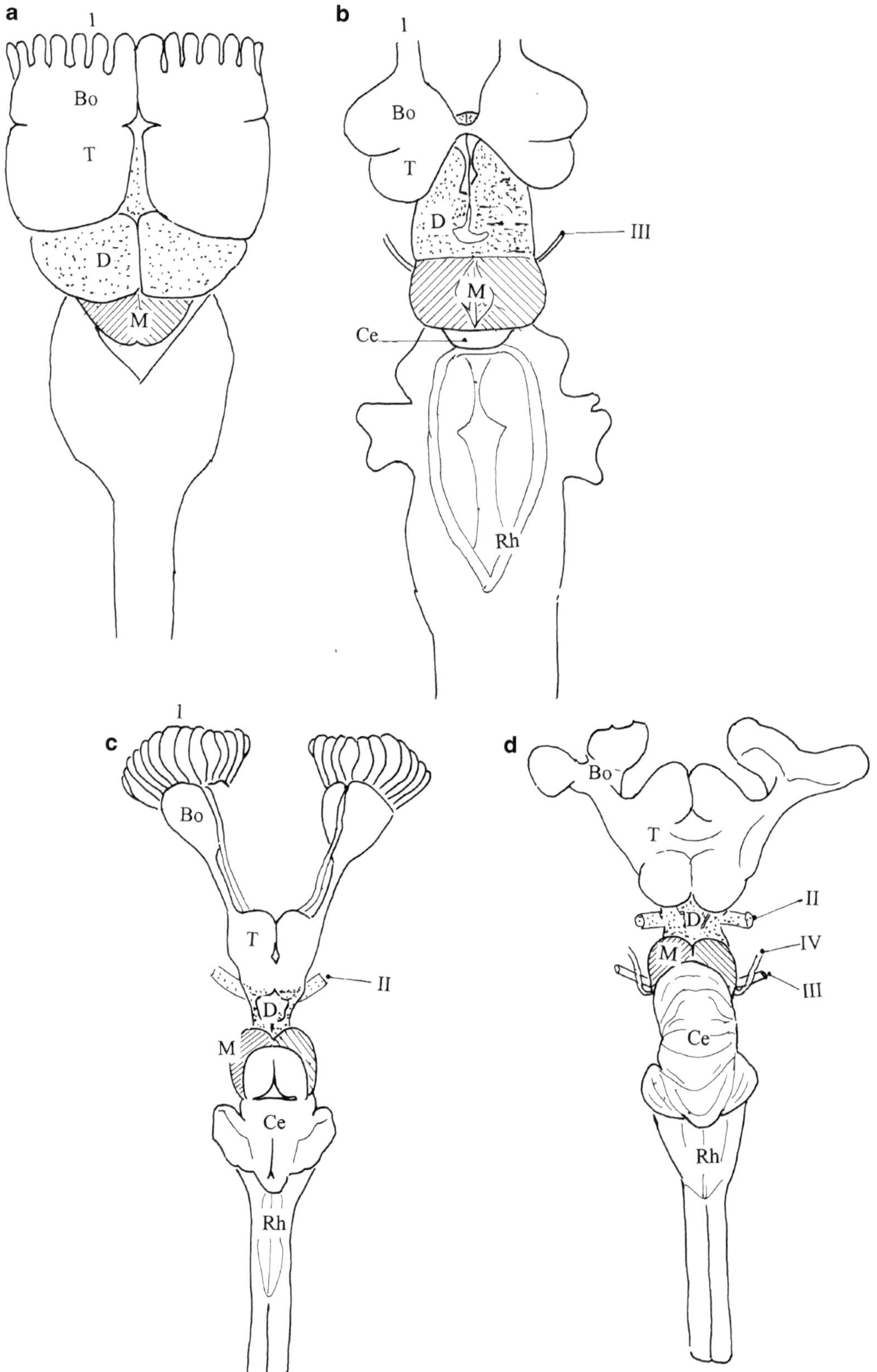

Fig. 3.2 *Variability of encephalon (development during ca. 800 million years) of some pisces* (according to Roth and Wullimann [9], p. 20; Indian ink copy with correction of cranial nerves by the author). **a**: *Eptatretus stouti* (Wicht and Northcutt (1998), cit. Roth and Wullimann [9], p. 20f., 31), **b**: Ichthymyzon (Ulinski (1983), cit. Roth and Wullimann [9], p. 20f., 31), **c**: Squalus (selachial brain), **d**: Mustelus (selachial brain) (B and C: Roth and Wullimann [9], p. 21), *bo* bulbus olfactorius, *T* telencephalon, *D* diencephalon, *M* mesencephalon, *Ce* cerebellum, *Rh* rhombencephalon, *Ce and Rh* metencephalon, (*I*) n. olfactorius, (*II*) n. opticus, (*III*) n. oculomotorius, (*IV*) n. trochlearis

3.2 Amphibia (Fig. 3.3)

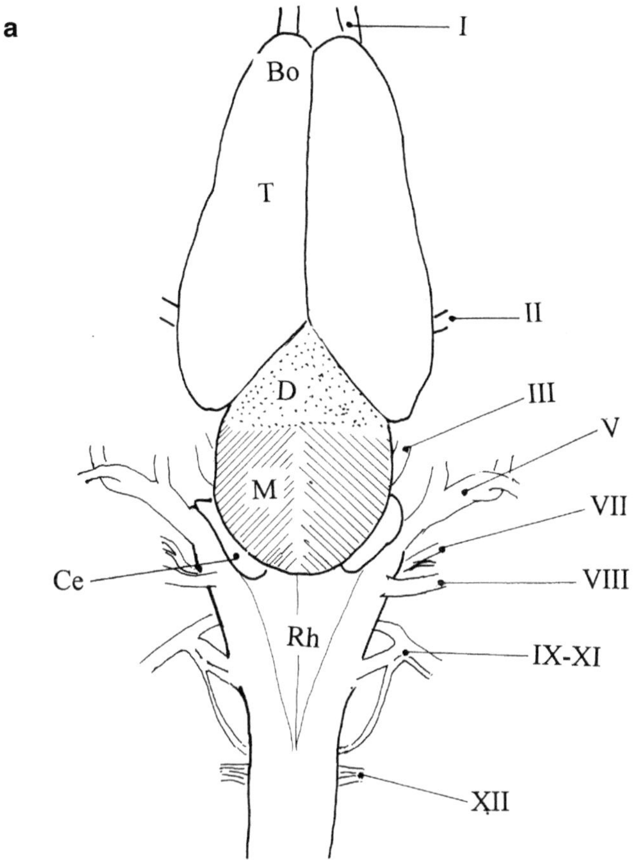

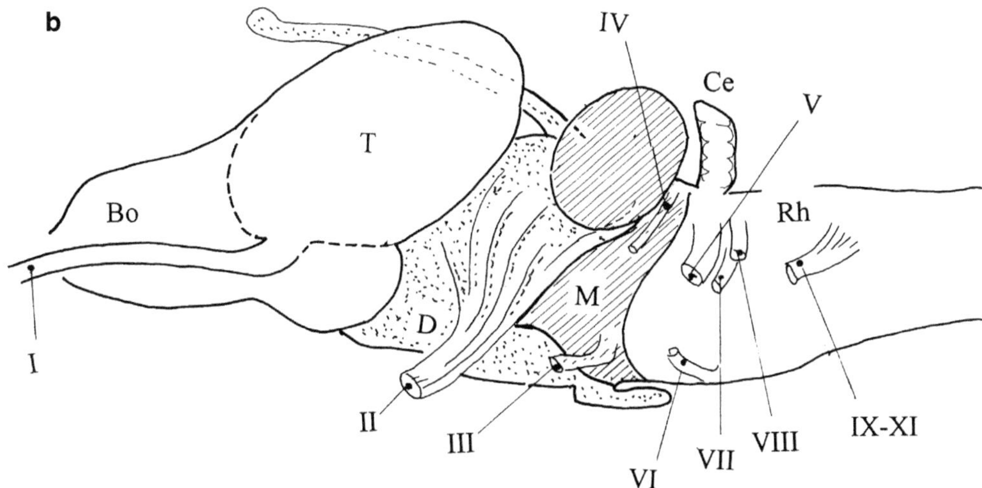

Fig. 3.3 *Branchial nerves of pisces are transformed to cranial nerves in all terrestric craniata. At* amphibia, n. VIII presents a transformation of a lateral ectodermal plate. This is a modification of the lateral ectodermal structures of pisces for controlling of stato-locomotoric functions. Amphibia have developed the first hearing of craniata. Most developed is hearing at mammalia, especially chiroptera (archaic Mammalia) and Cetacea (high developed Mammalia). Archaic cranial nerves nn. I, II, III, IV, VI, and XII of pisces remained. (**a**) Triton hydromantes, *Bo* bulbus olfactorius, *T* telencephalon, *D* diencephalon, *M* mesencephalon, *Ce* cerebellum, *Rh* rhombencephalon (Ce and Rh: metencephalon). (**b**) Rana

3.3 Reptilia and Aves (Both Together: Sauropsides) (Fig. 3.4)

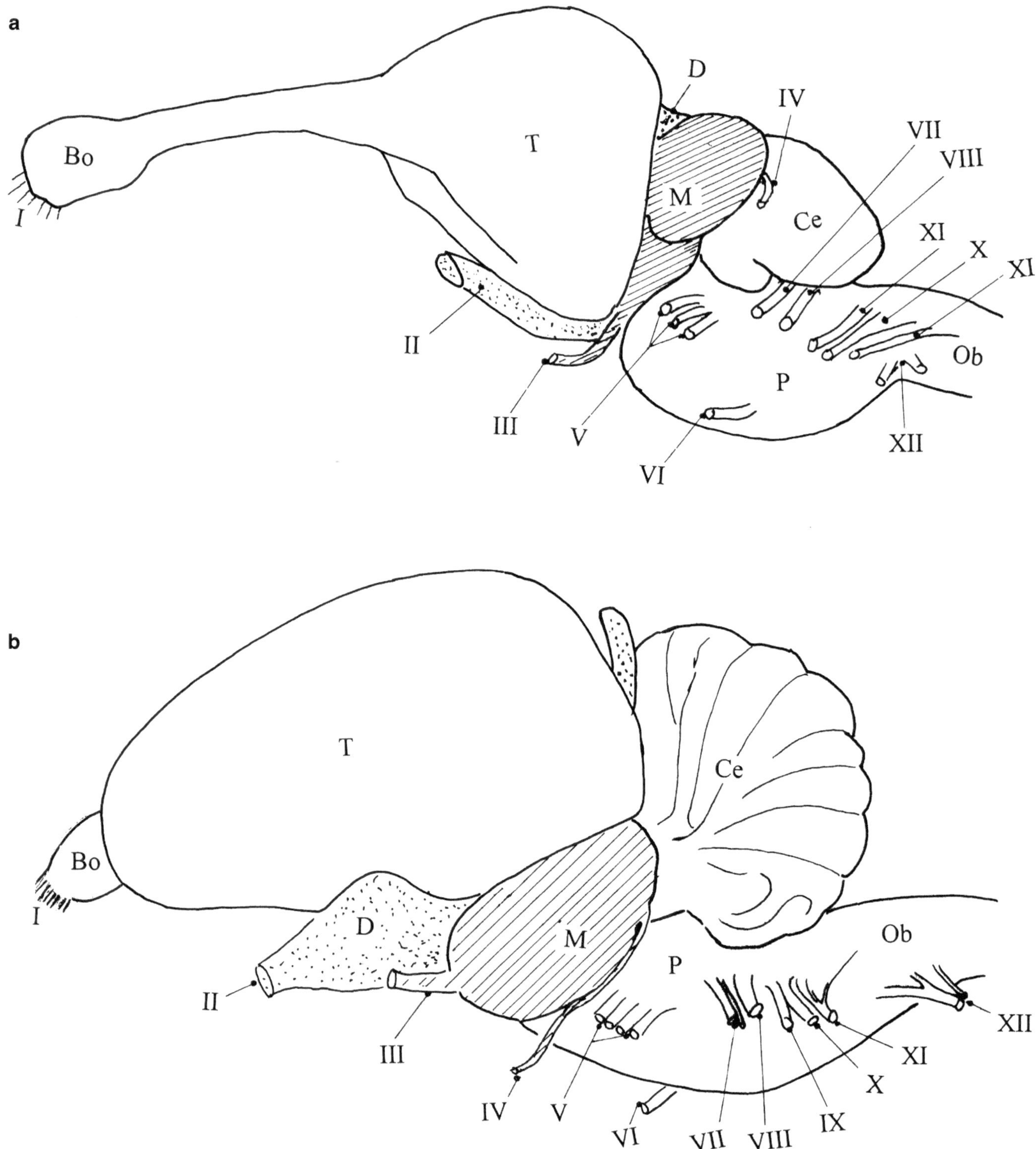

Fig. 3.4 (**a**) *Reptile (alligator)*, (**b**) *avis (goose)*, *Bo* bulbus olfactorius, *T* telencephalon, *D* diencephalon, *M* mesencephalon, *Ce* cerebellum, *P* pons, *Ob* medulla oblongata

3.4 **Mammalia** (Figs. 3.5, 3.6, 3.7, 3.8, 3.9, and 3.10)

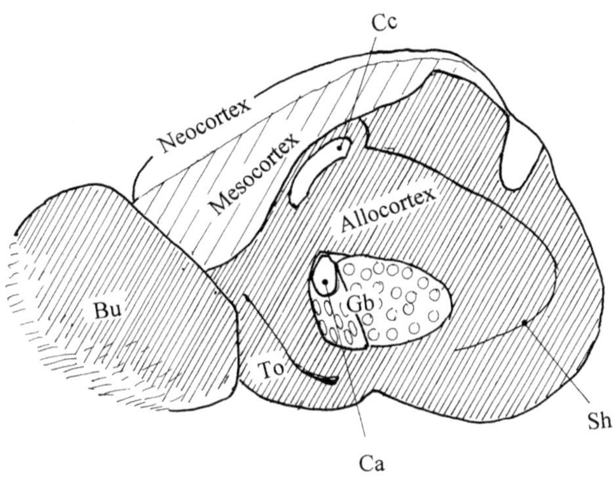

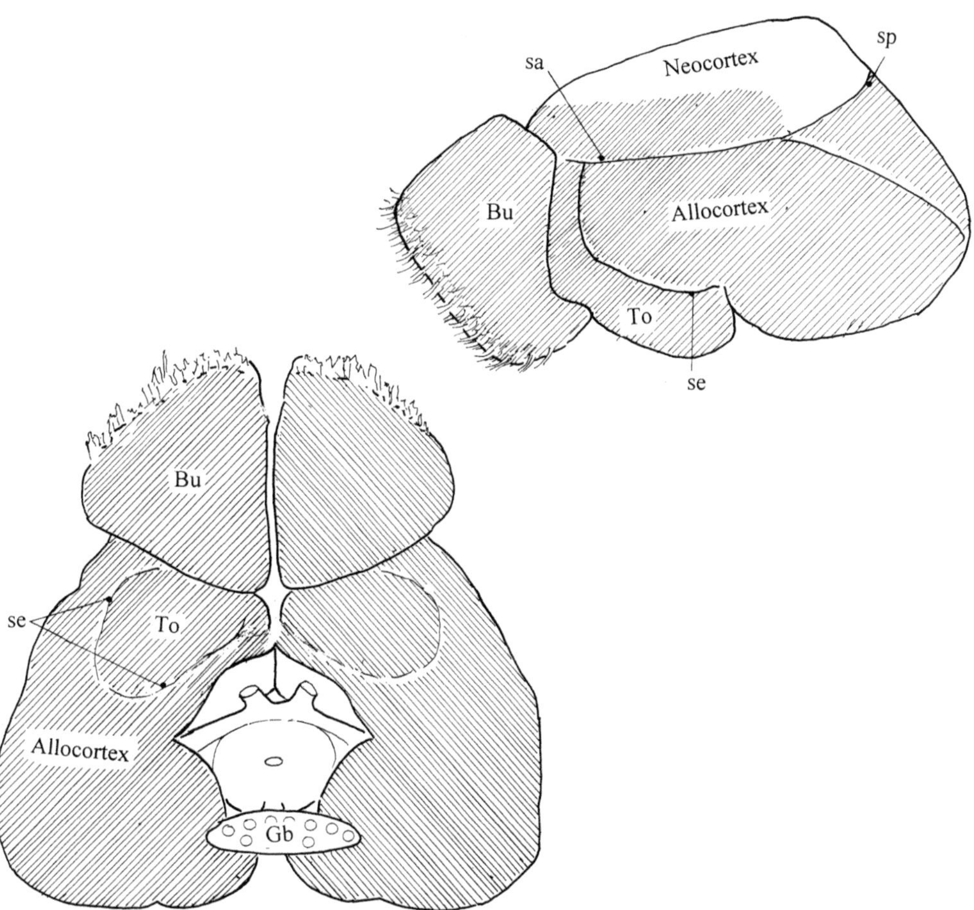

Fig. 3.5 *Small neocortex of an archaic mammal*: *Erynaceus europaeus* (hedgehog) (according to Stephan [13], p. 30ff). Hatching: Allocortex and mesocortex. Mesocortex: Mixtum of allocortex and neocortex. *Bu* bulbus olfactorius, *To* tuberculum olfactorium, *Ca* commissura anterior, *Gb* gangliae basales, *Sh* sulcus hippocampalis, *Ce* corpus callosum, *se* sulcus entorhinalis, *sp* sulcus rhinalis posterior, *sa* sulcus rhinalis anterior

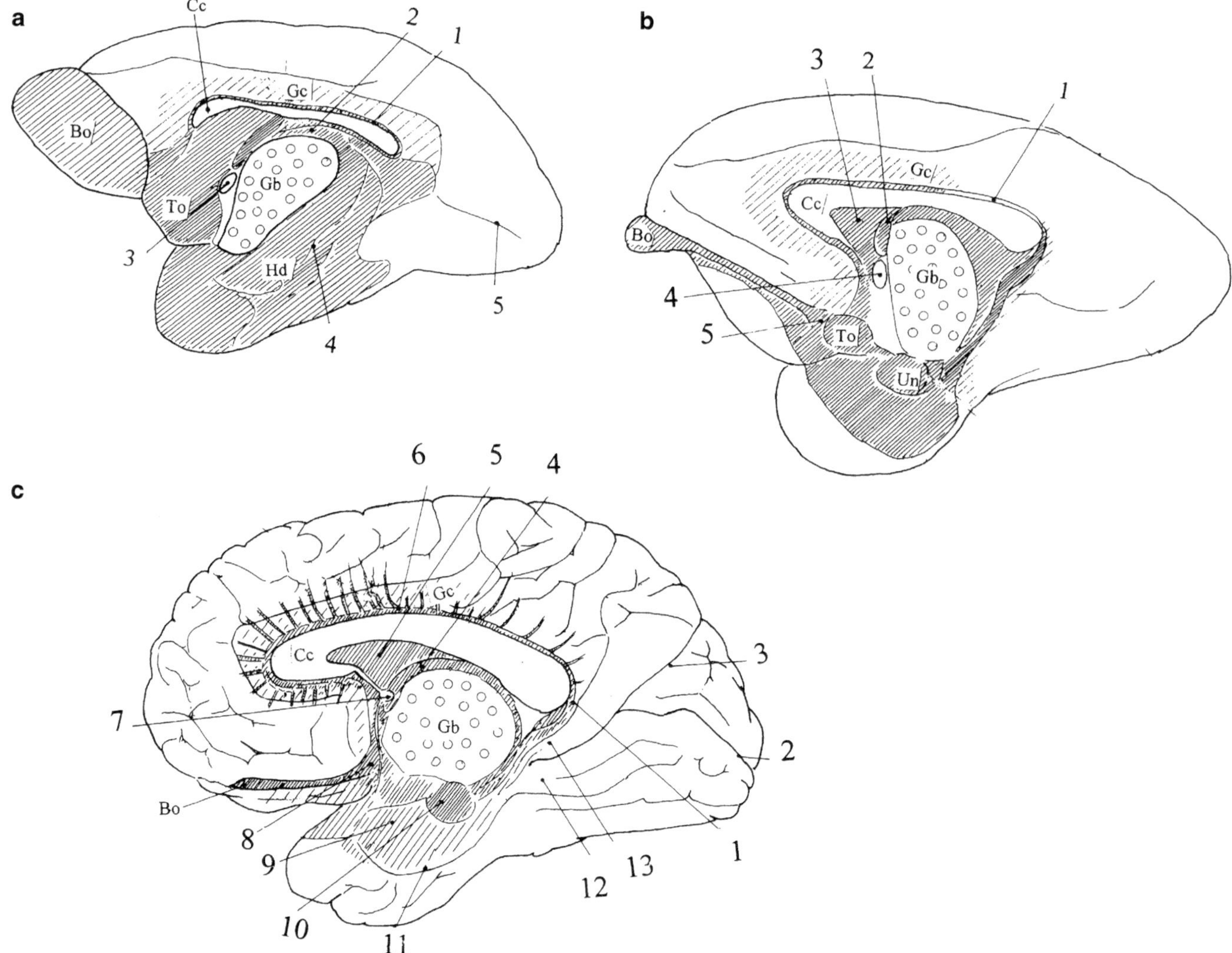

Fig. 3.6 *Early apadides and homo.* **a**: Prosimia (*Galago demidovi*) (Stephan [13], see Fig. 3.5, p. 44, modified Indian ink copy of the author). Commissura anterior and corpus callosum small. **b**: Simia (*Cercopithecus ascanius*). Increase of corpus callosum and commissura anterior. **c**: Homo. Increase of corpus callosum and reduction of commissura anterior (Stephan [13], see Fig. 3.5). **a**: (*1*) splenium corporis callosi, (*2*) septum pellucidum, (*3*) commissura anterior, (*4*) hippocampus–dentatus–parahippocampalis–complex. **b**: (*1*) splenium corporis callosi, (*2*) fornix, (*3*) septum pellucidum, (*4*) commissura anterior, (*5*) trigonum olfactorium. **c**: (According to fiber dissections of the author (see Figs. 4.3 and 4.5)) (*1*) stria longitudinalis cinguli, (*2*) sulcus calcarinus, (*3*) sulcus parietooccipitalis, (*4*) fornix, (*5*) septum pellucidum, (*6*) as (*1*), (*7*) commissura anterior, (*8*) trigonum olfactorium, (*9*) gyrus parahippocampalis, (*10*) as (*9*), (*11*) sulcus occipitotemporalis medialis, (*12*) gyrus occipitotemporalis lateralis, (*13*) gyrus occipitotemporalis medialis. **a–c**: *Bo* bulbus olfactorius, *Cc* corpus callosum, *Gb* gangliae basales, *Gc* gyrus cinguli, *Hd* gyrus hippocampus–dentatus-area, *To* tuberculum olfactorium. Dark hatching: allocortex, light hatching: mesocortex (allo- and mesocortex mixed)

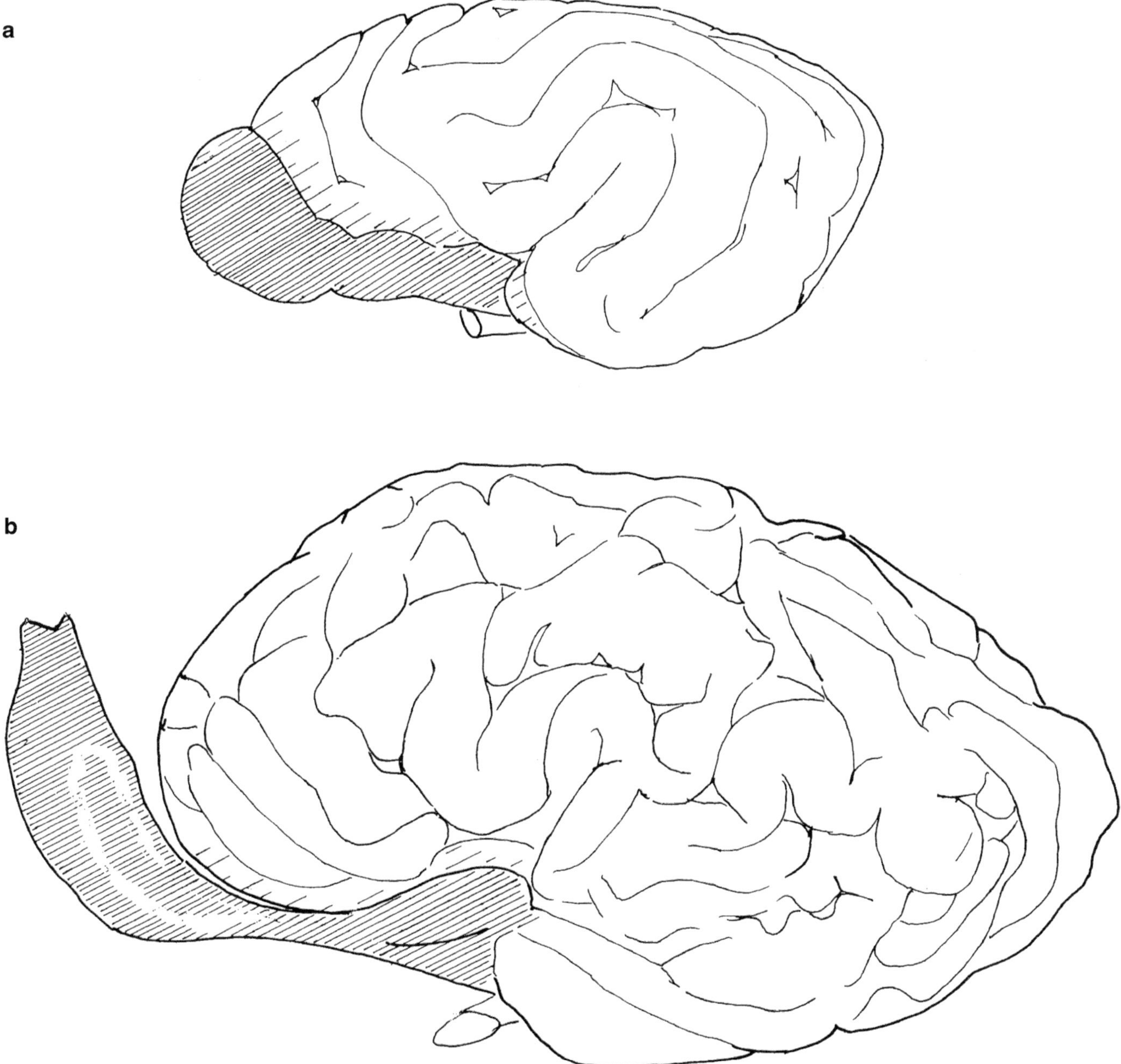

Fig. 3.7 *Well-gyrificated mammals with well-developed rhinencephalic part of allocortex (hatched)*. (**a**) Canis domesticus, (**b**) equus

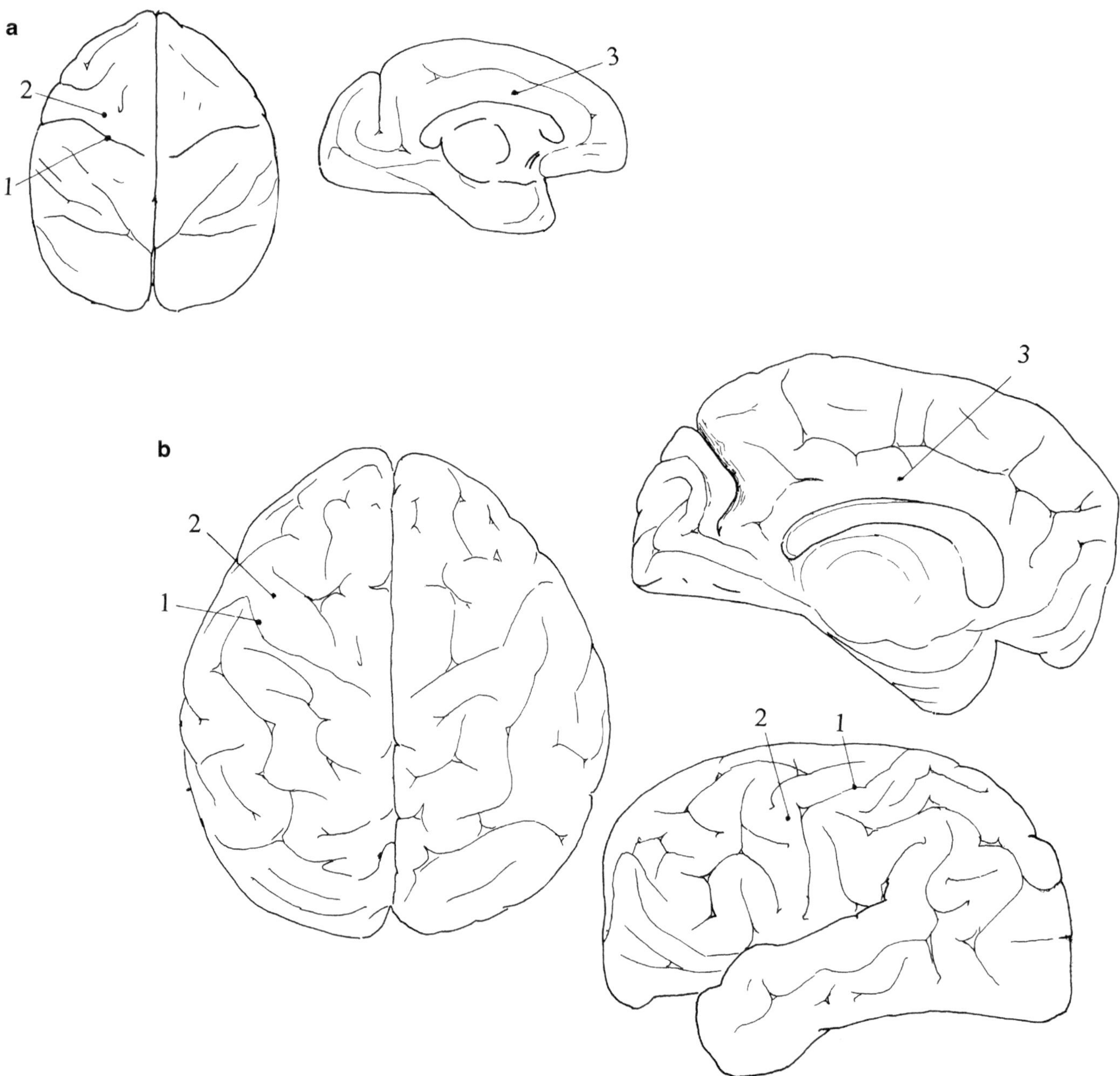

Fig. 3.8 *Early apadide and early hominide* (von Bischoff [15], Tafel VI and VII). **a**: Macacus, **b**: orang utan, (*1*) sulcus centralis, (*2*) gyrus praecentralis, (*3*) gyrus cinguli

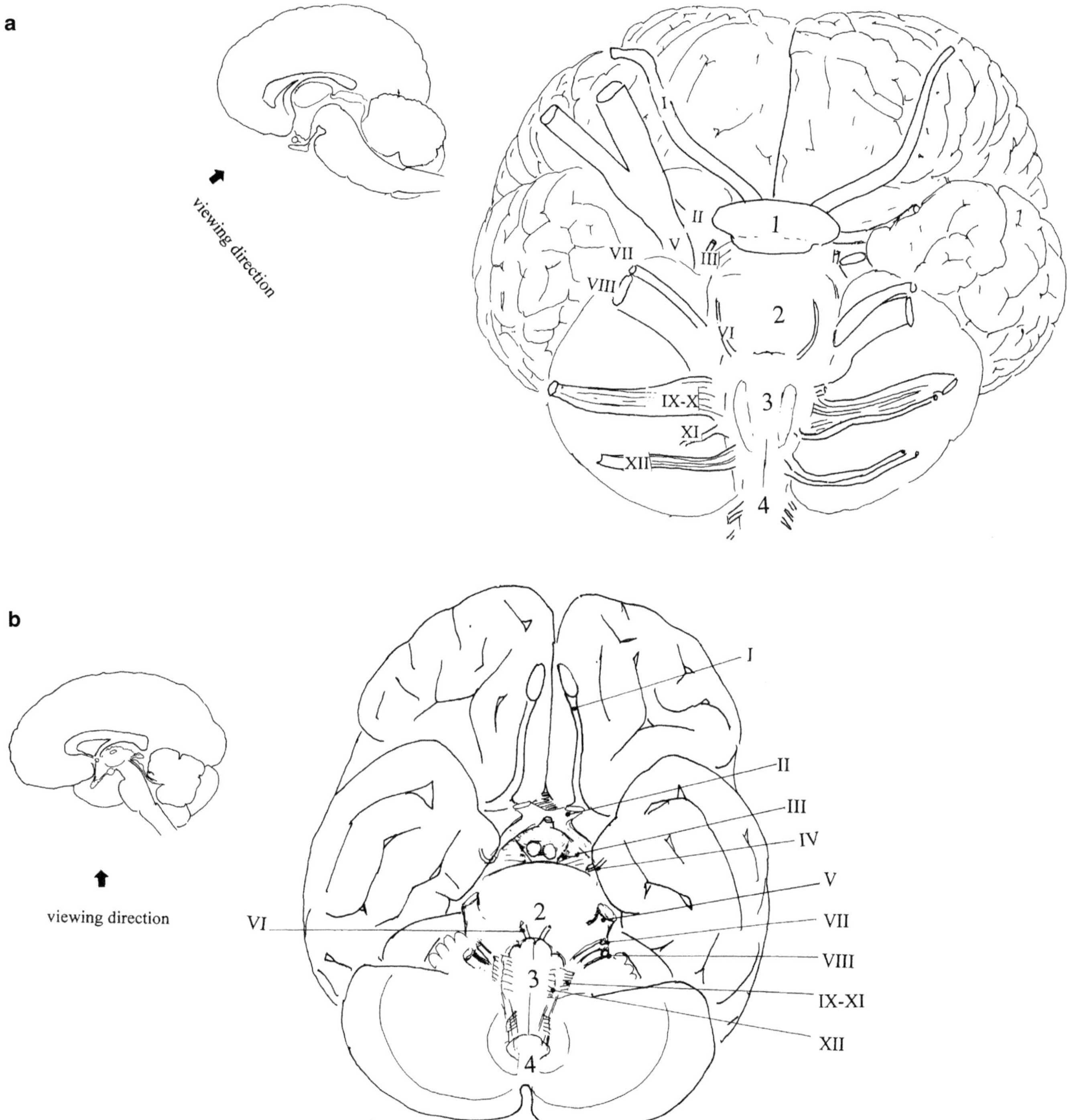

Fig. 3.9 *Dolphin* (Pilleri, Giehr, and Kraus [5], pp. 373–388. Indian ink copy of the author, simplified). **a**: *and homo*, **b**: Gyrification is more developed at dolphin than at homo. Cranial nerves of dolphin are adapted to the larger mass of body. Allocortex omitted. (*1*) Hypophysis, (*2*) pons, (*3*) medulla oblongata, (*4*) medulla spinalis

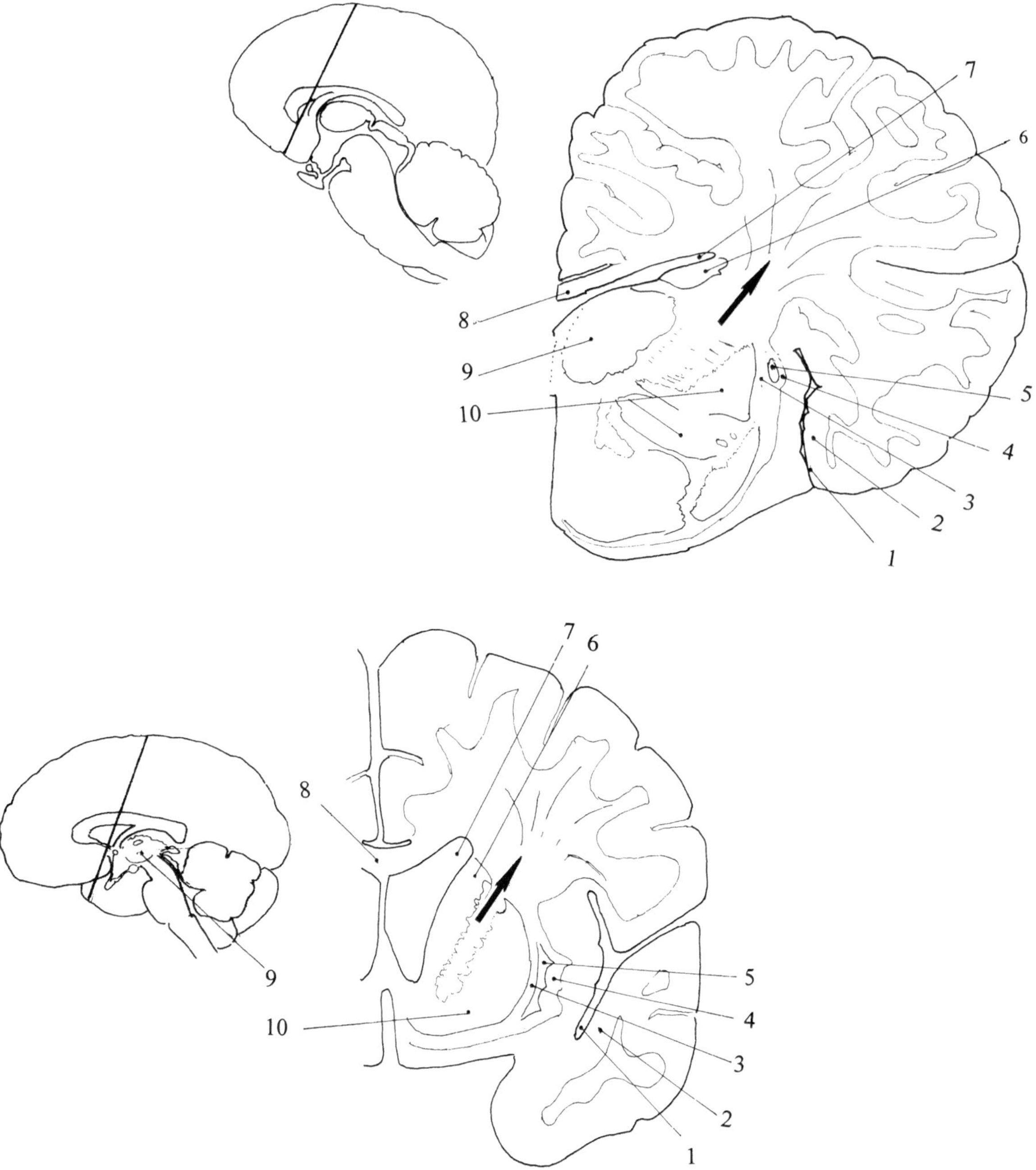

Fig. 3.10 *Continuation. Transectional planes.* Gyrification of dolphin is more developed than at homo. Gyri are smaller. The lateral ventricle and corpus callosum of dolphin are small. (*1*) Sulcus lateralis (Sylvii), (*2*) insula, (*3*) capsula externa, (*4*) capsula extrema, (*5*) claustrum, (*6*) nucleus caudatus, (*7*) ventriculus lateralis, (*8*) corpus callosum, (*9*) thalamus, (*10*) putamen et pallidum. Arrow: Beginning of corona radiata and capsula interna

Résumé of Figs. 3.4–3.10 The high decreased allocortex of aves presents a new kind of telencephalic development.

A new kind of telencephalic development of mammalia is the first presentation of the stepwise high magnificated neocortex, combined with the reduction of allocortex (in homo less than 10% of the weight of telencephalon[22]). All allocortex and neocortex are closely connected to each other. Consciousness is not possible without an intact allocortex. Consciousness may remain intact after extirpation of neocortex of the nondominant frontal lobe, if all bilateral allo- and mesocortical elements and fronto-parieto-occipito-temporal are preserved. Exact transection of the interhemispheric corpus callosum is often tolerated without following functional deficits. Defects of corpus callosum may be compensated by commissura anterior and posterior. Exact callosotomy is very difficult to perform without lesion of stria longitudinalis cinguli. Stria longitudinalis cinguli is located 4 mm lateral from the midline. Lesions of this stria (and gyrus cinguli) may be followed by severe psychological deficits[23]

[22] (Stephan [13], p. 138, Abb. 130–131)

[23] (Seeger [11], p. 167 ff)

Part II

Exceptional Position of the Human Encephalon

4 Telencephalon, Survey

See Figs. 4.1, 4.2, 4.3, 4.4, and 4.5.

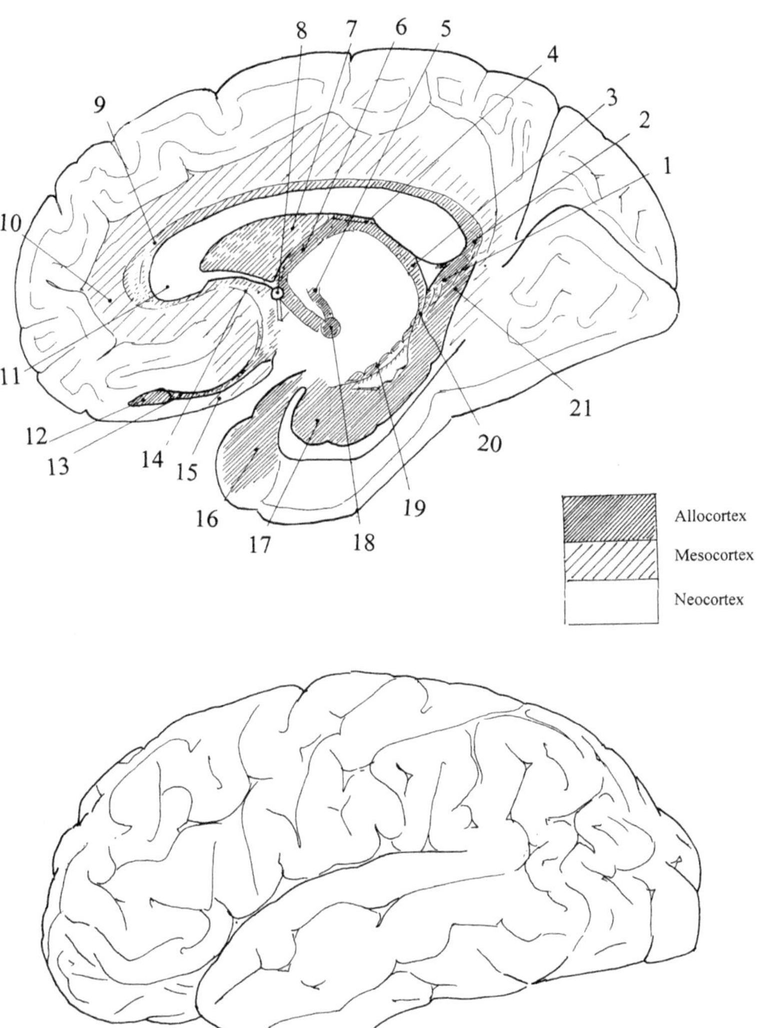

Fig. 4.1 *Allocortex and neocortex*. Connections of allocortex with diencephalon (corpus amygdyloides with basal gangliae and tractus mamillothalamicus). (*1*) Posterior dorsal continuation of area dentata, (*2*) beginning of indusium griseum and stria longitudinalis cinguli, (*3*) crus fornicis, (*4*) commisura fornicis, (*5*) tractus mamillothalamicus, (*6*) columna fornicis, (*7*) septum pellucidum, (*8*) commisura anterior, (*9*) stria longitudinalis cinguli, (*10*) gyrus cinguli, (*11*) genu corporis callosi, (*12*) bulbus olfactorius, (*13*) tractus olfactorius, (*14*) area subcallosa, (*15*) gyrus rectus, (*16*) amygdala, (*17*) pes hippocampi, (*18*) corpus mamillare, (*19*) gyrus dentatus, (*20*) finbria fornicis, (*21*) extraventricular tail of hippocampus. Mesocortex: mixtum of allocortex and neocortex (light dotted)

W. Seeger, *Evolution of the Central Nervous System of Craniata and Homo*, https://doi.org/10.1007/978-3-030-15216-1_4

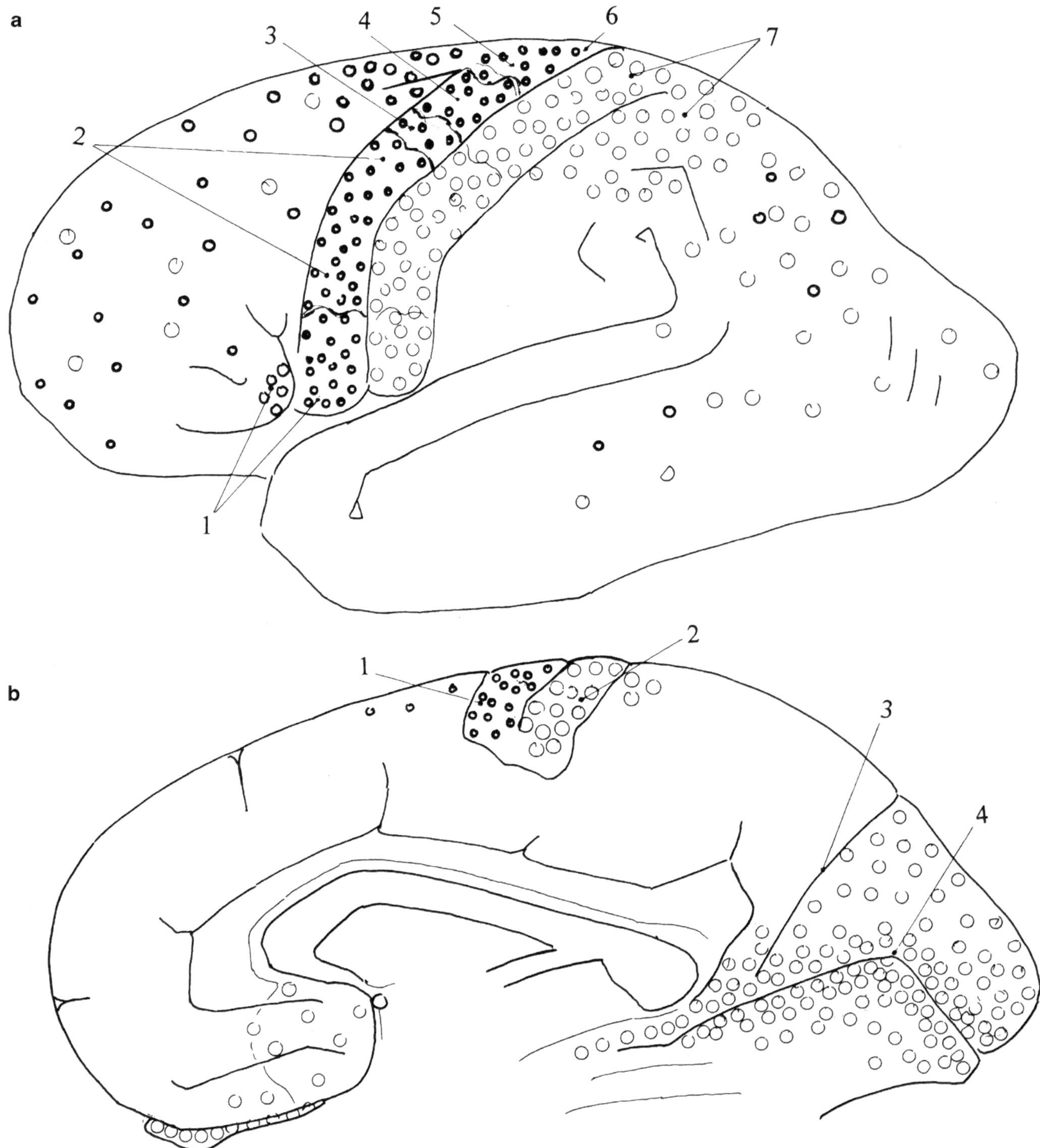

Fig. 4.2 *Neocortical centers* (individual variability proved by neurophysiological methods) *of the left dominant hemisphere. For centers of further special psychological functions* see Fig. 10.2. **a**: lateral. (*1*) one of the right- or left-sided located motor centers for speech (Broca), (*1*) and (*2*) frontal extrapyramidal system and variable motor speech centers, (*2*) motor center for the right face, the right hand, (*3*) of the right arm, (*4*) of the trunk, (*5*) of the bladder, (*6*) of both legs, (*7*) sensoric area for the right side. **b**: medial. (*1*) Gyrus paracentralis, anterior segment, motor center for bladder and both legs, (*2*) gyrus paracentralis, posterior segment, sensoric centers, (*3*) sulcus parietooccipitalis, superior limit area of the primary visual center, (*4*) sulcus calcarinus, central visual center. Between parietal area and posterior segment of gyrus parahippocampalis: secondary optic region

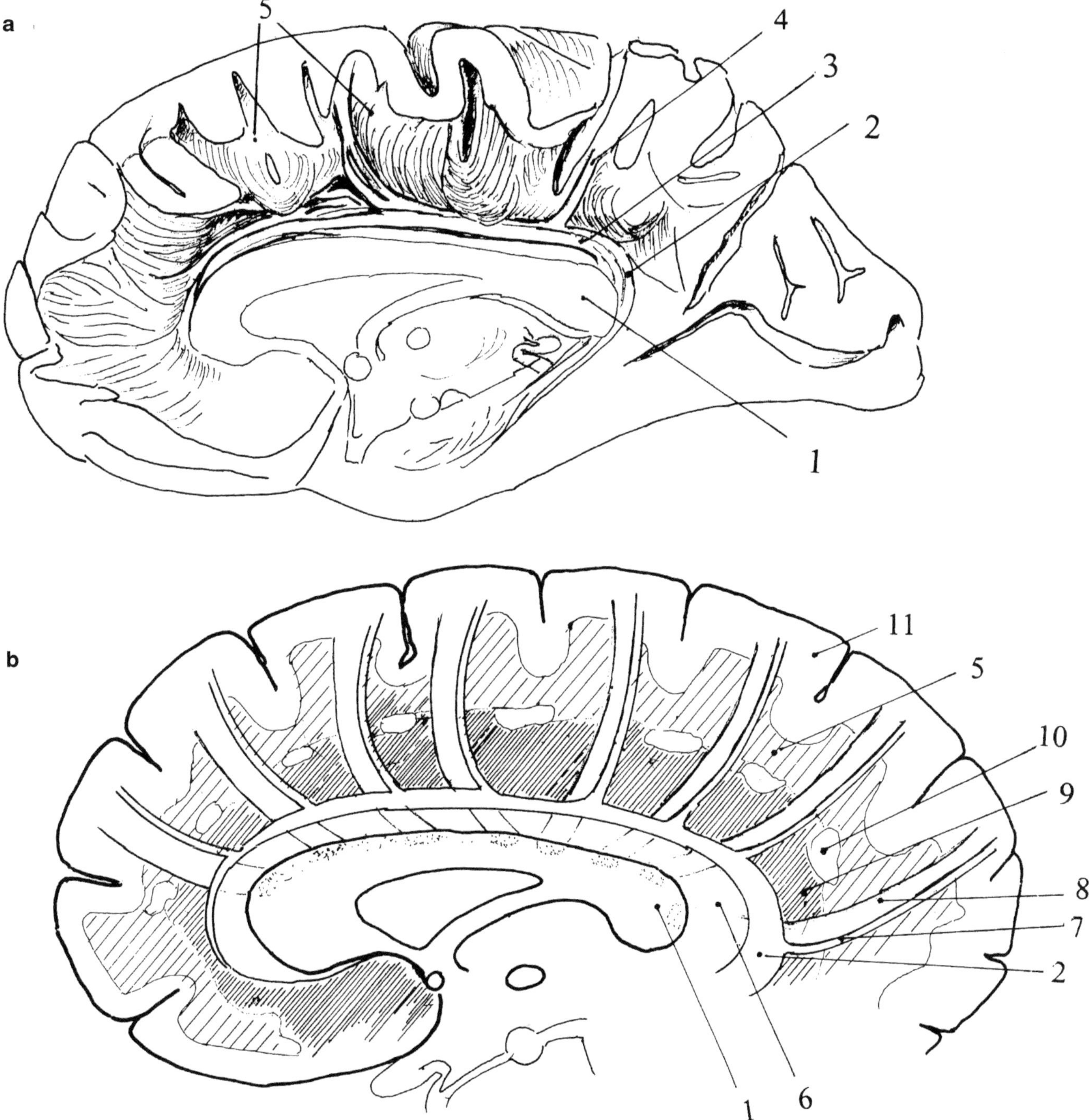

Fig. 4.3 *Allocortex and neocortex with systematic presentation of its fibers*. Anatomical fiber dissection (**a**) and schematized presentation (**b**). (*1*) Splenium corporis callosi, (*2*) continuation of area dentata into 3, (*3*) stria longitudinalis cinguli, (*4*) fiber bundles of corpus callosum, (*5*) U-fibers, (*6*) dorsal surface of corpus callosum, (*7*) fiber bundles of stria longitudinalis cinguli, (*8*) fiber bundles of corpus callosum, (*9*) gyrus cinguli, (*10*) fascicle(s) of corona radiata, (*11*) cortex

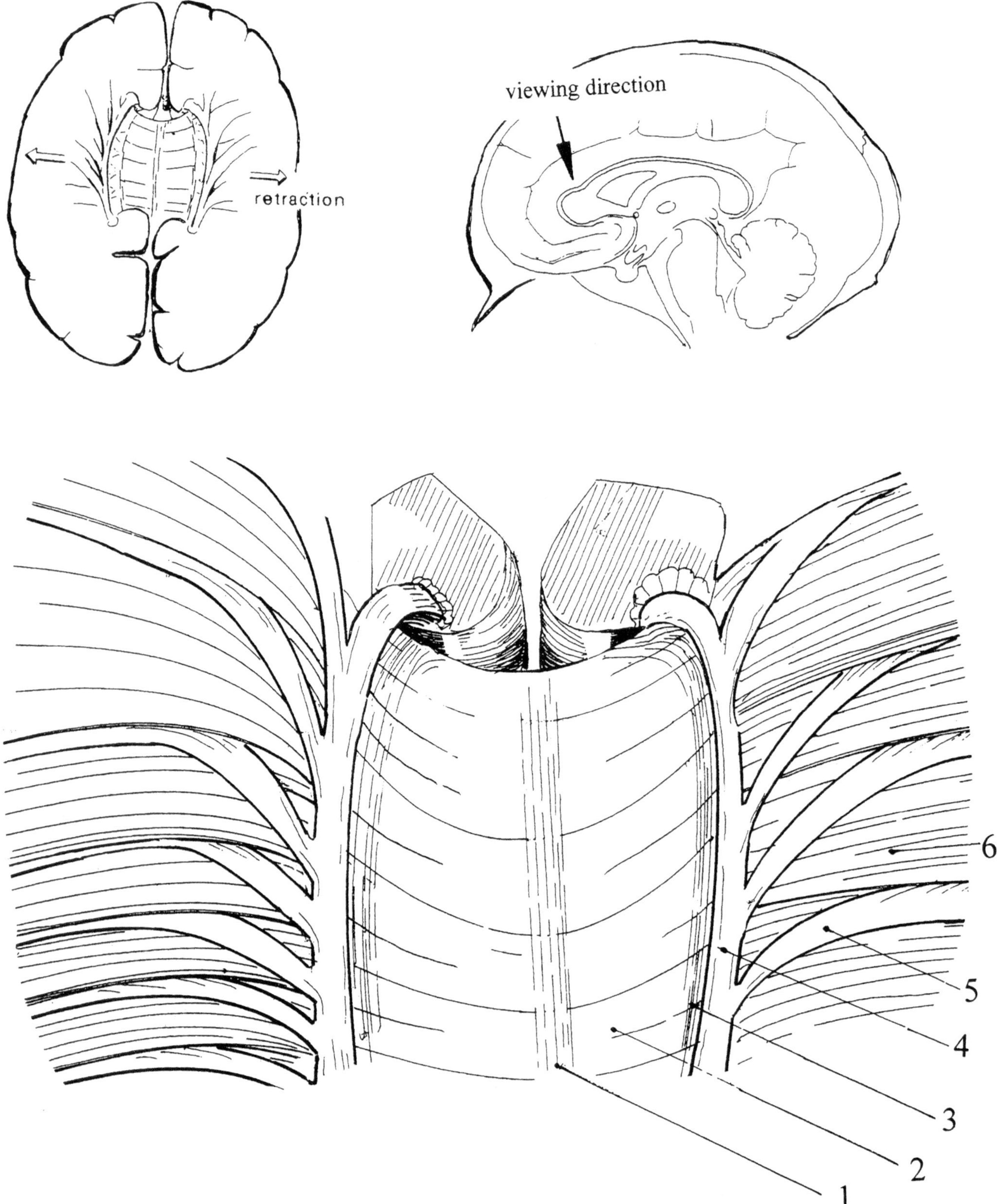

Fig. 4.4 (Seeger [11], p. 166f) *Anatomical dissection of fiber bundles of corpus callosum and stria lateralis cinguli.* (*1*) Indusium griseum (striae mediales), (*2*) corpus callosum, (*3*) indusium griseum (striae laterales), (*4*) stria longitudinalis cinguli, (*5*) fiber bundle of (*4*), (*6*) fiber bundles of corpus callosum

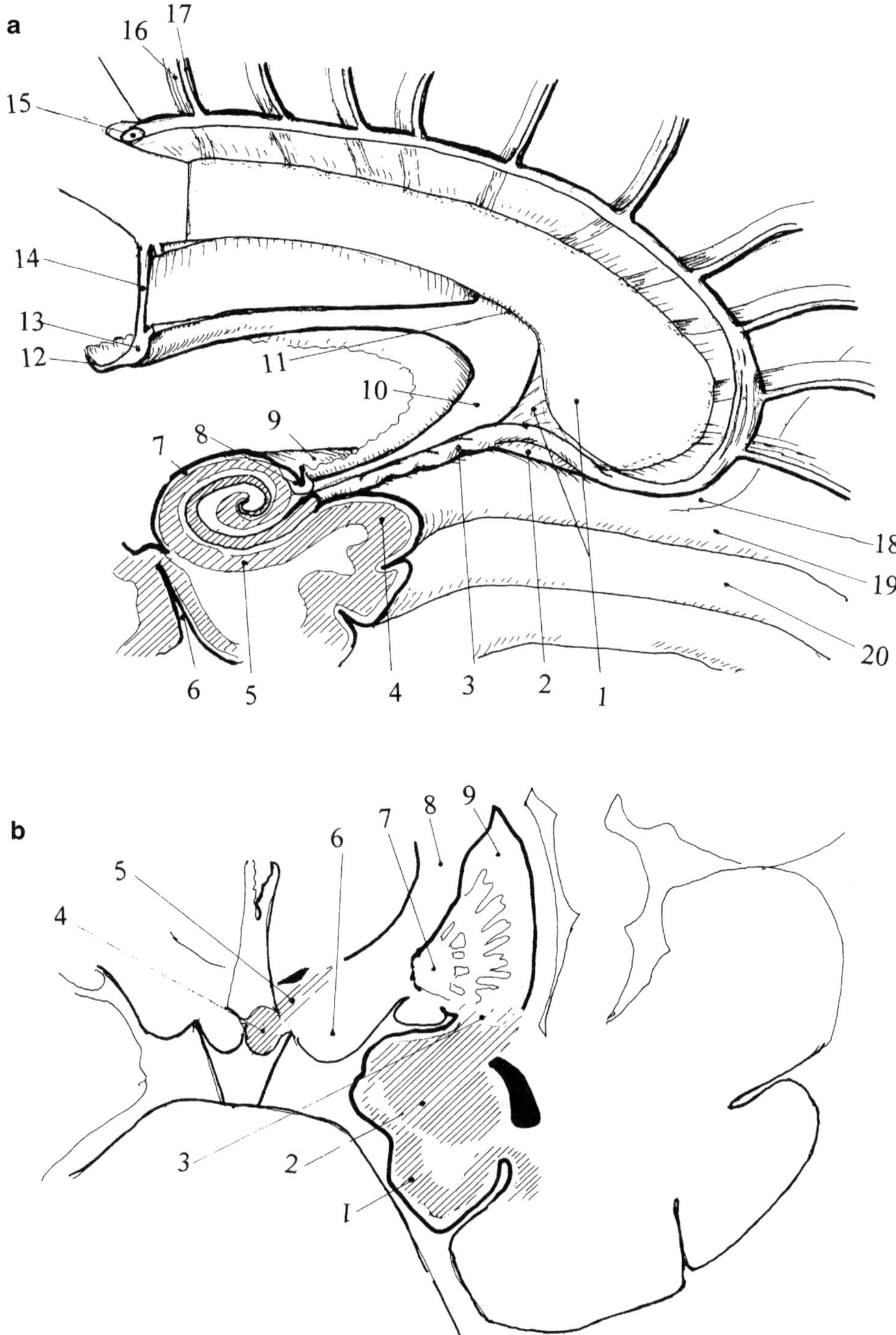

Fig. 4.5 *Allocortex dorsal from corpus callosum and fornical connections with diencephalon.* **a**: allocortex dorsal. (*1*) splenium corporis callosi, (*2*) extraventricular part of hippocampus (tail of h), (*3*) gyrus dentatus, (*4*) gyrus parahippocampalis, (*5*) hippocampus, (*6*) sulcus collateralis, (*7*) alveus (fibers of fornix), (*8*) fimbria fornicis, (*9*) as (*7*), (*10*) crus fornicis, (*11*) commissura fornicis, (*12*) taenia fornicis, (*13*) corpus/columna fornicis, (*14*) septum pellucidum, (*15*) stria longitudinalis cinguli, (*16*) fiber bundle of corpus callosum, (*17*) fiber bundle of stria longitudinalis cinguli, (*18*) isthmus gyri cinguli, (*19*) gyrus occipitotemporalis medialis, (*20*) gyrus occipitotemporalis lateralis. **b**: fornical connection with diencephalon and upper brainstem (tegmentum of mesencephalon and upper pons region). (*1*) gyrus uncinatus, (*2*) amygdala, (*3*) *allocortical-diencephalic connections* which are connected with brainstem[1], (*4*) corpus mamillare (basis of columna fornicis), (*5*) tractus mamillothalamicus, *allocortical-diencephalic fiber connection*, (*6*) crus cerebri, (*7*) pallidum, (*8*) capsula interna, (*9*) putamen. *Telencephalic and cerebellar centers are not so exactly structured than other old cerebral structures* (Figs. 5.1, 5.2, and 5.3)

[1] Ad 3 and 5 in **b**: Circumscribed lesions of formatio reticularis of mesencephalon and upper pons region are followed by unconsciousness and tonic seizures by touching of extremities (well known by neurosurgeons and neurologists), the so-called decerebration. Respiratory functions and further vegetative functions are preserved, but modified. (Seeger [10]).

5 Diencephalon, Mesencephalon, Rhombencephalon, Cerebellum: Survey

See Figs. 5.1, 5.2, and 5.3.

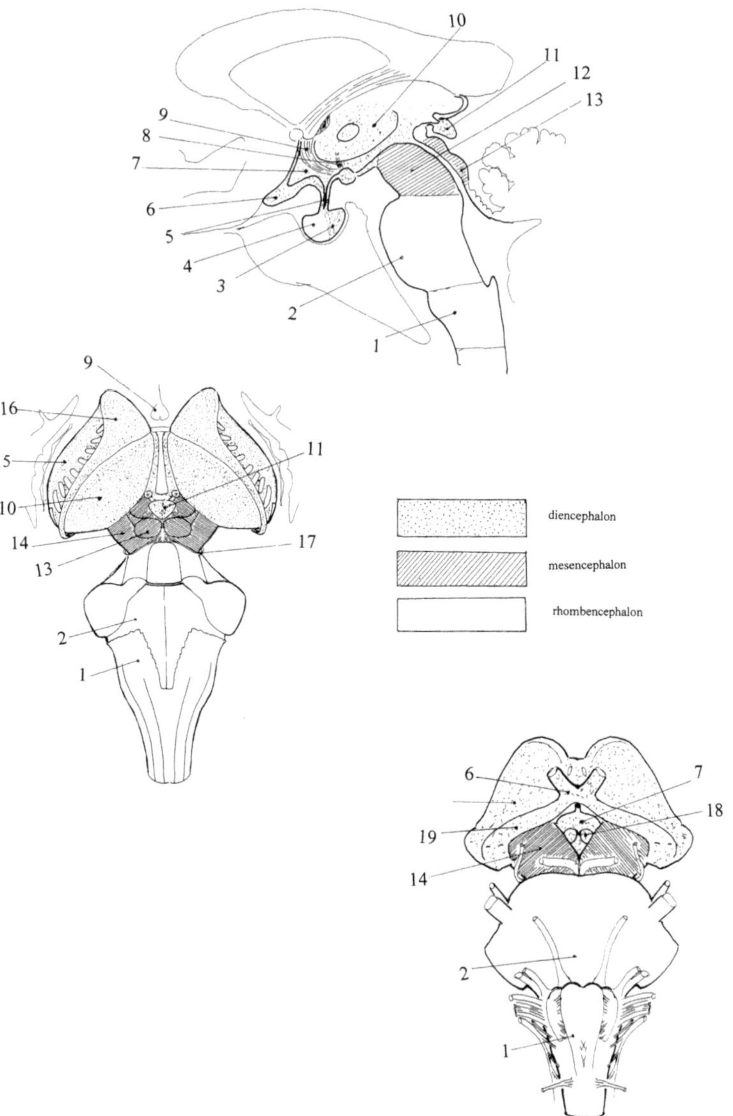

Fig. 5.1 *Diencephalon, mesencephalon, and rhombencephalon of homo. High variability in craniata. In craniatas all these parts of encephalon are presented exactly symmetric and systematic, and corpus callosum of telencephalon of mammalia and homo included.* (*1*) Medulla oblongata, (*2*) pons, (*3*) neurohypophysis, (*4*) adenohypophysis, (*5*) infundibulum, (*6*) chiasma, (*7*) hypothalamus, (*8*) tractus mamillothalamicus, (*9*) columna fornicis, (*10*) thalamus, (*11*) corpus pineale, (*12*) tegmentum mesencephali, (*13*) tectum mesencephali, (*14*) crus cerebri, (*15*) putamen-pallidum, (*16*) nucleus caudatus, (*17*) isthmus rhombencephali, (*18*) corpus mamillare, (*19*) tractus opticus

W. Seeger, *Evolution of the Central Nervous System of Craniata and Homo*, https://doi.org/10.1007/978-3-030-15216-1_5

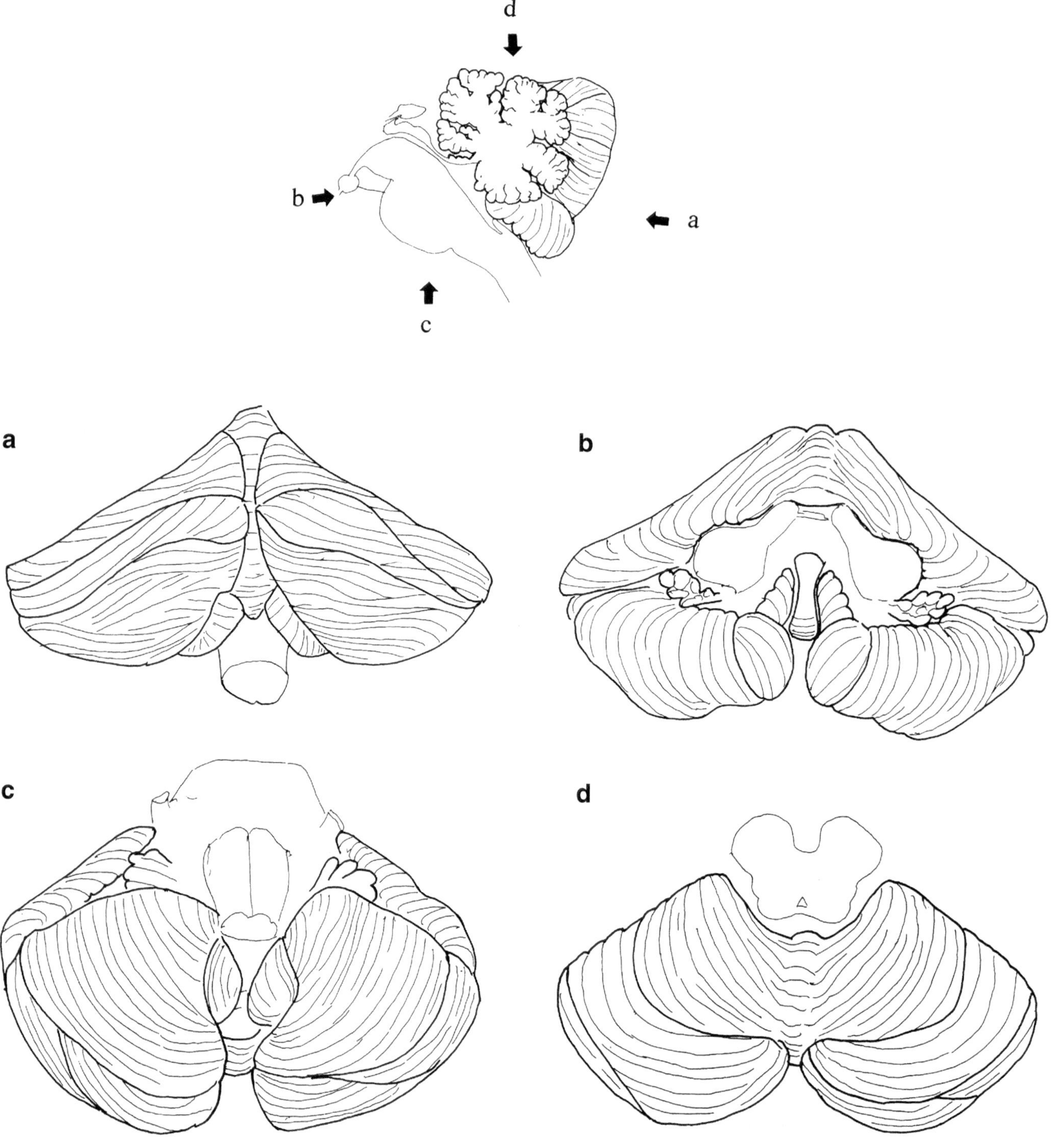

Fig. 5.2 *Cerebellum*. Bilateral asymmetry of its hemispheres (neocerebellum). Predominant symmetry of vermis and flocculus (palaeocerebellum)

Fig. 5.3 *Systematic alternating three types of fiber plates* (**a–c**): (According to Lang. Personal fiber dissections of Prof. J. Lang, Würzburg. Here fiber dissections by the author), (**c**) schematized. (*1*) Fiber plates of pedunculus cerebellaris superior, (*2*) of pedunculus cerebellaris medius, (*3*) nucleus dentatus, (*4*) cerebello-dentato fiber plates

6 Telencephalon: Details

6.1 Definition of Telencephalic Fiber Systems and Gyri: Introduction

During the progredient increasing of neocortex, its cortical components and underlying fiber bundles were folding into gyri and U-fibers. In principle, U-fibers are archaic associative fibers. These archaic fibers are the predominant fibers of telencephalon. Younger fiber systems are the long associative fiber fascicles of the white matter of hemispheres (centrum semiovale). Long associative fiber fascicles are presented small in all mammals. Long connections are predominant between frontal lobe and other lobes. The intralobal frontal between distant gyri is missed, although lobus frontalis is the largest telencephalic lobus. Only at occipital lobus are located some longer associative fiber connections. Large long fiber connections of centrum semiovale present the projection systems. These are fibers of corpus callosum and fibers of corona radiata. The U-fibers with roundabout connections are predominant and not the long direct associative fiber fascicles between the telencephalic lobes. These are immature structures in contrast to projection fiber systems.

W. Seeger, *Evolution of the Central Nervous System of Craniata and Homo*, https://doi.org/10.1007/978-3-030-15216-1_6

6.1.1 Gyrification and U-Fibers (Figs. 6.1, 6.2, 6.3, 6.4, and 6.5)

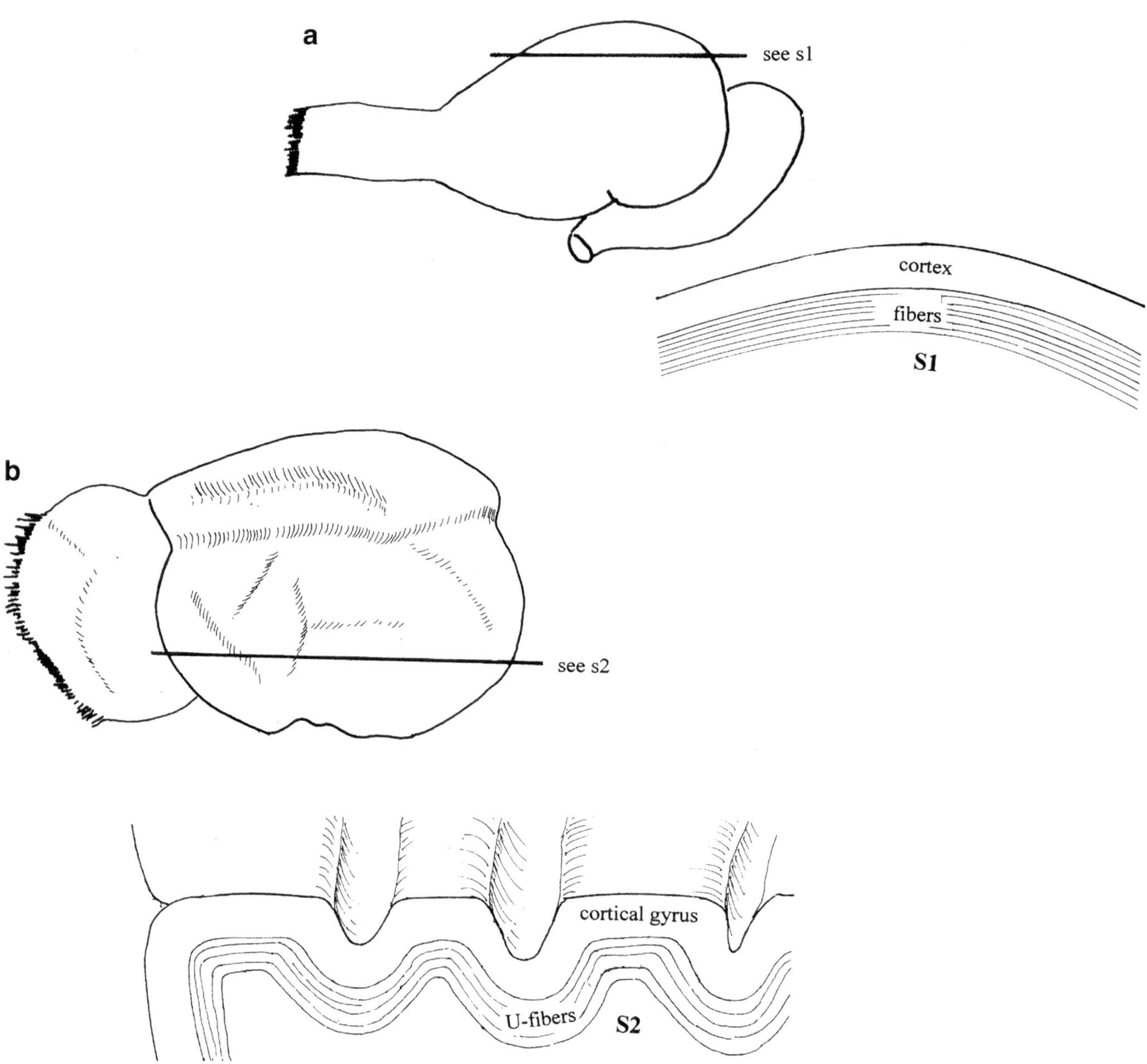

Fig. 6.1 *Agyria and beginning gyrification* (Stephan [13], p. 21. Simplified presentation). (**a**) Agyria of a reptile (testudo, simplified presentation). (**b**) Beginning gyrification of an archaic mammal (*Erynaceus europaeus*, simplified presentation). *Subgyral fibers are the beginning of roundabout U-fibers* (Fig. 6.2)

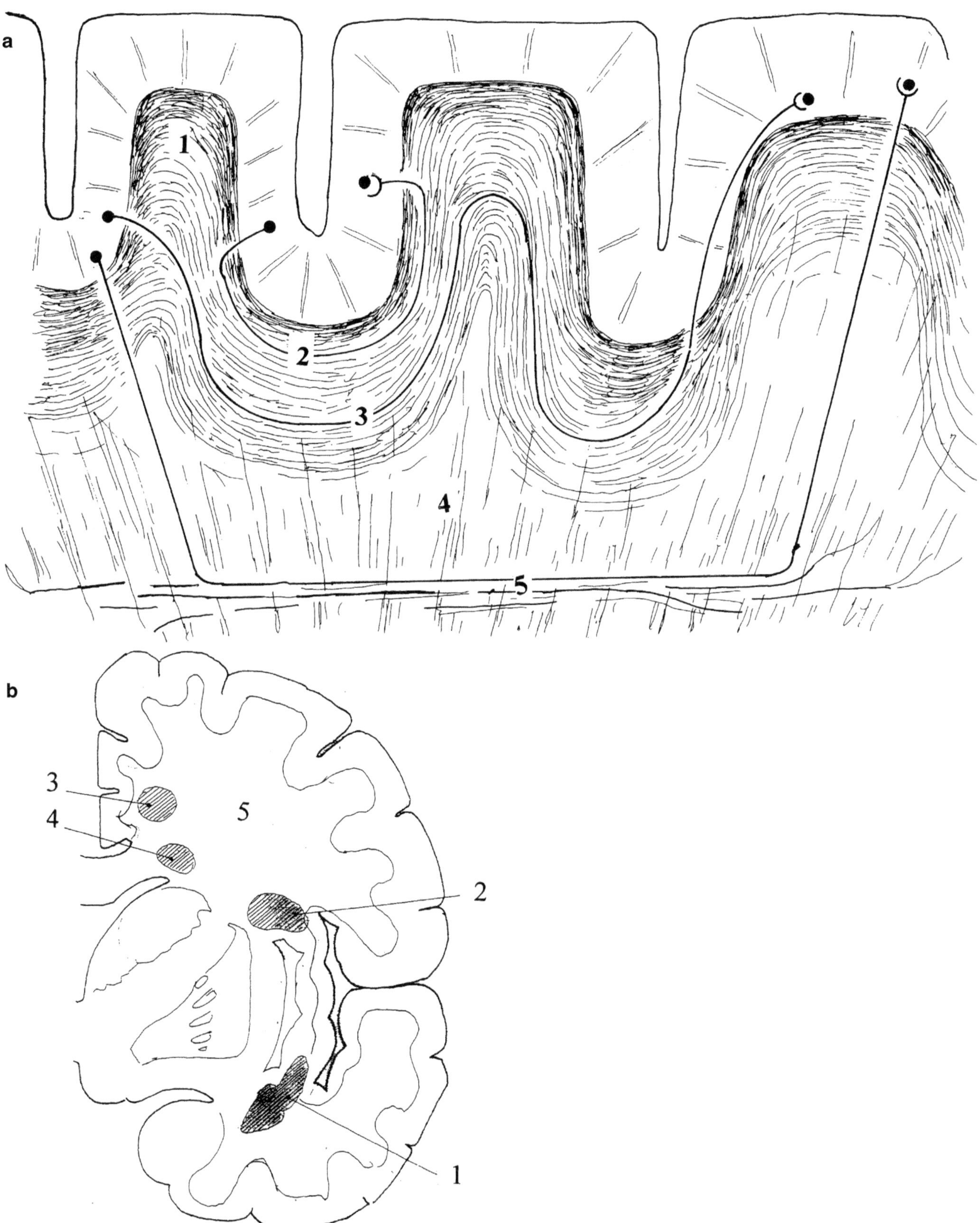

Fig. 6.2 *U-fibers* (**a**) *and centrum semiovale* (**b**). *U-fibers are an archaic association system of mammalian neocortex with roundabout fibers. It is the predominant association system in mammalia and not the younger long association fascicles* (**b**). **a**: (*1*) U-fibers, (*2*) short connections, (*3*) roundabout long connections of U-fibers, (*4*) predominant projection fibers of centrum semiovale, (*5*) association fascicle of centrum semiovale. **b**: association fascicles (Kahle and Frotscher [3], p. 261), (*1*) fasciculus occipitofrontalis, (*2*) fasciculus longitudinalis superior, (*3*) fasciculis cinguli, (*4*) fasciculus subcallosus, (*5*) centrum semiovale

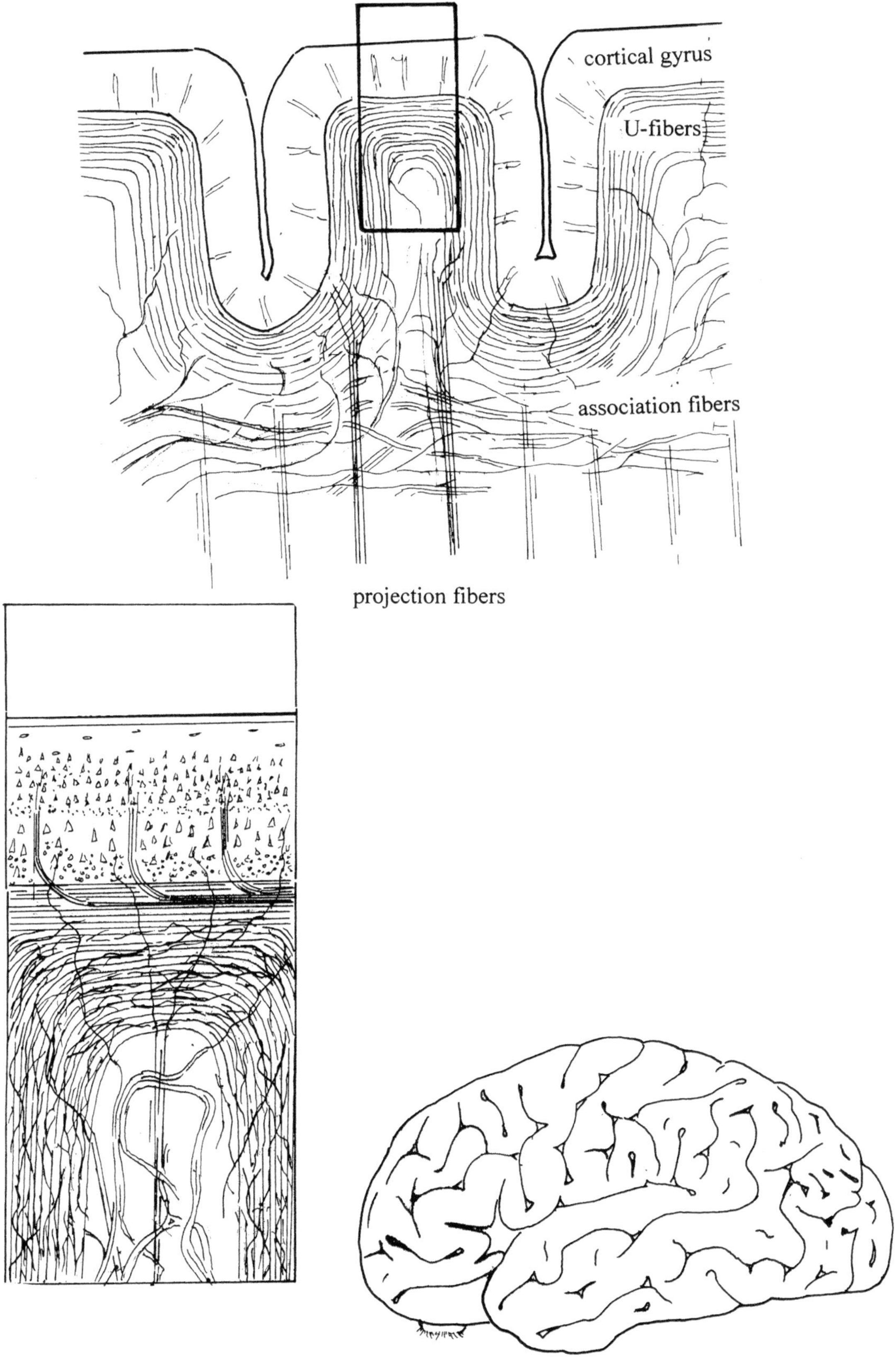

Fig. 6.3 *Addendum*

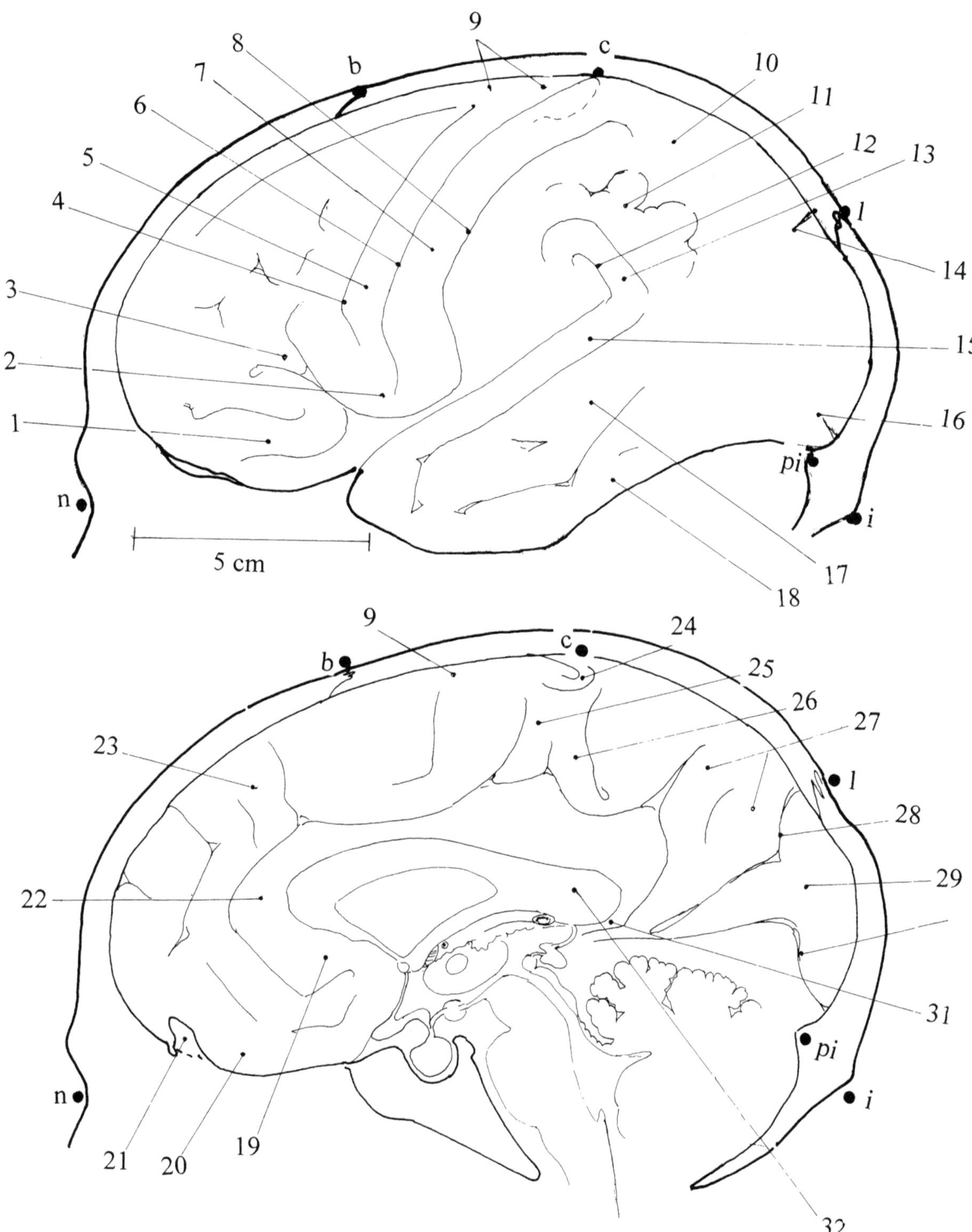

Figs. 6.4 and 6.5 *Topography of gyri. Landmarks: n* nasion, *b* bregma, *c* sulcus centralis, upper point, *l* lambda, *i* inion, *pi* protuberantia occipitalis interna. *Cerebral structures*: (*1*) third gyrus frontalis, pars orbitalis, (*2*) pars opercularis, (*3*) pars triangularis, (*4*) sulcus centralis, (*5*) gyrus praecentralis, (*6*) sulcus centralis, (*7*) gyrus postcentralis, (*8*) sulcus postcentralis, (*9*) precentral motor planning region, (*10*) upper parietal region, (*11*) gyrus circumflexus, (*12*) sulcus lateralis (Sylvii), pars ascendens, (*13*) gyrus angularis (Wernicke), (*14*) sulcus parietooccipitalis, dorsal point, (*15*) first gyrus temporalis, (*16*) sulcus calcarinus, posterior point, (*17*) second gyrus temporalis, (*18*) third gyrus temporals, (*19*) area subcallosa, (*20*) gyrus rectus, (*21*) crista galli, (*22*) gyrus cinguli, (*23*) first frontal gyrus, (*24*) sulcus centralis, upper point, (*25*) gyrus paracentralis, anterior segment, (*26*) posterior segment, (*27*) praecuneus, (*28*) sulcus parietooccipitalis, (*29*) cuneus, (*30*) sulcus calcarinus, (31) isthmus gyri cinguli, (*32*) splenium corporis callosi, (*33*) (Fig. 6.5) bulbus olfactorius, (*34*) tractus olfactorius, (*35*) gyrus occipitotemporalis lateralis, (*36*) Uncus, (*37*) gyrus parahippocampalis and gyrus occipitotemporalis medialis

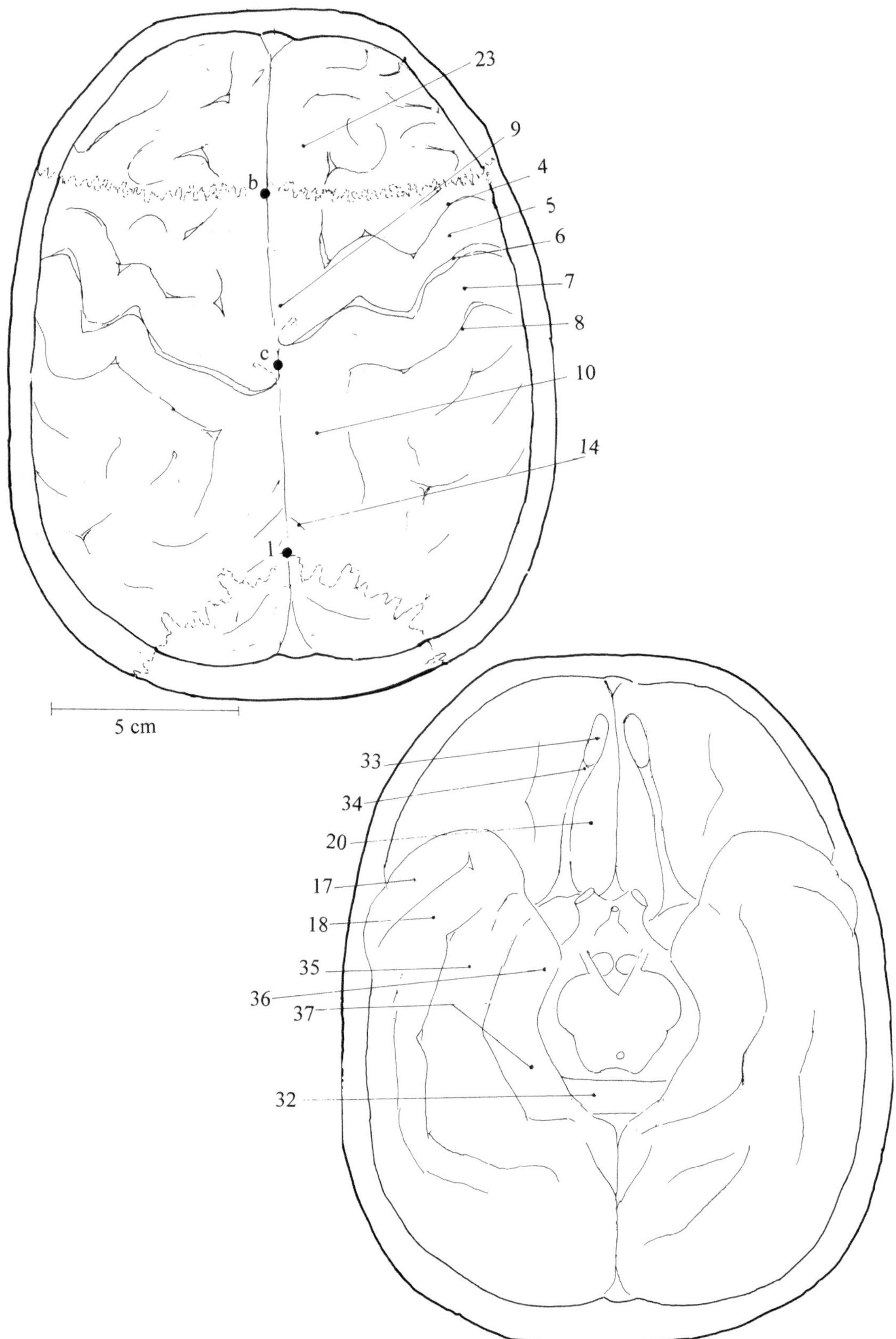

Fig. 6.5 Addendum for Fig. 6.4. *Dorsal and basal gyri*

6.1.2 **Association Fibers** (Figs. 6.6, 6.7, and 6.8): **Anatomical Fiber Dissections**

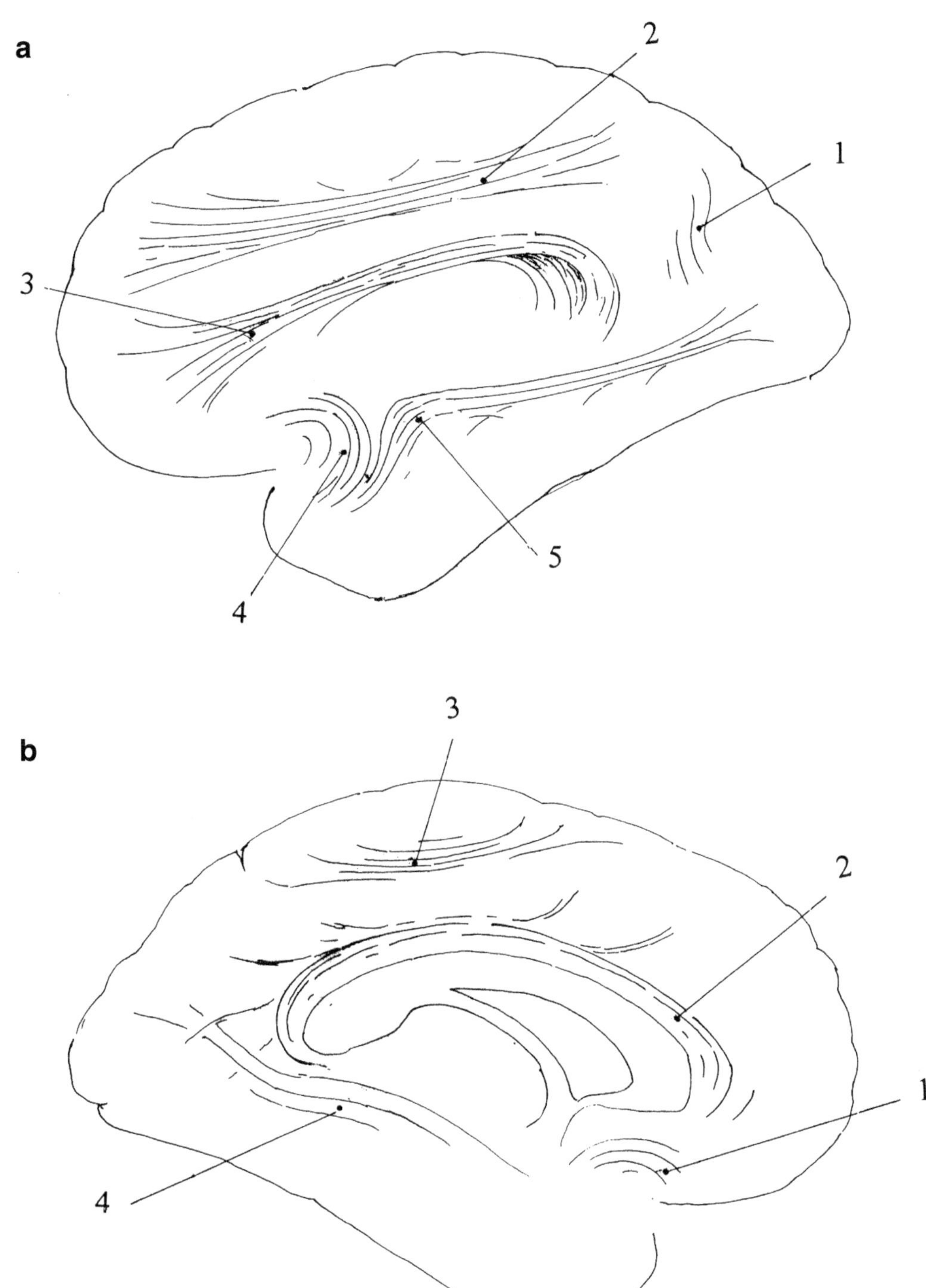

Fig. 6.6 *Association fascicles of centrum semiovale* (Ludwig and Klingler [4]. Modified Indian ink copies of the author) *and of gyrus cinguli.* **a**: lateral: (*1*) fasciculi verticales, (*2*) fasciculus longitudinalis superior, (*3*) as (*2*), (*4*) fasciculus uncinatus, (*5*) fasciculus occipito-frontalis; **b**: medial: (*1*) fasciculus uncinatus, (*2*) fasciculus cingularis, (*3*) fasciculus frontoparietalis, (*4*) fasciculus occipitotemporalis. Most long connections are connections of the frontal regions with parietal, occipital, and temporal regions. Only the small long connections lateral from area are intraoccipital connections. There are no intrafrontal long connections within the wide frontal white matter. Long intrafrontal connections are only archaic roundabout frontal U-fibers which are mixed with other fibers

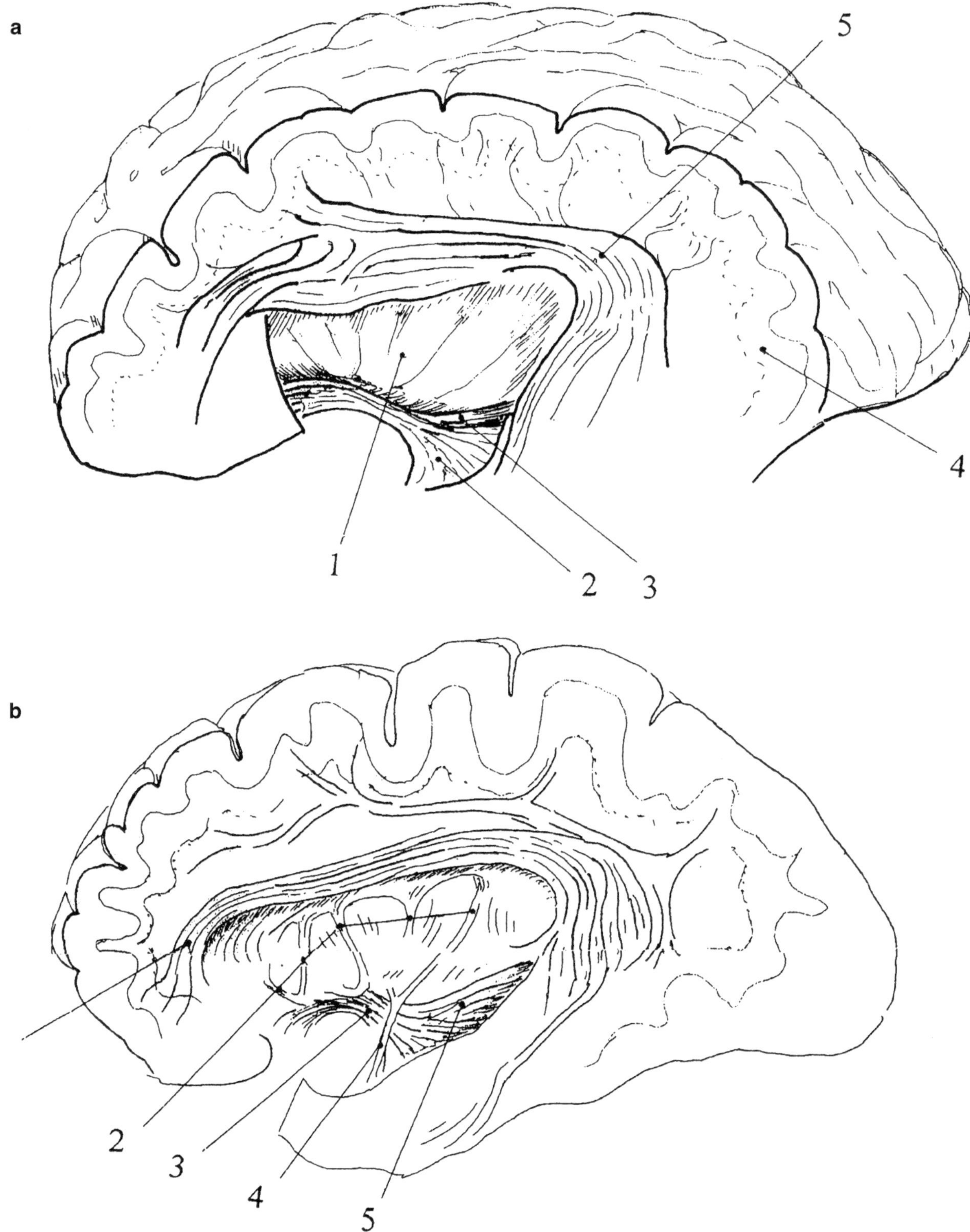

Fig. 6.7 *Continuation*. (**a**) Fiber dissection dorsolateral and basomedial from Fig. 6.6a. (*1*) Insula, (*2*) fasciculus uncinatus, (*3*) fasciculus occipitofrontalis, (*4*) U-fibers and centrum semiovale, (*5*) fasciculus frontoparietotemporalis. (**b**) Dorsal medial oblique and basal lateral fiber dissection from (**a**): (*1*) fasciculus frontoparietotemporalis, (*2*) cortical fibers of insula, (*3*) fasciculus uncinatus, (*4*) fiber connections with uncus, (*5*) fasciculus occipitofrontalis

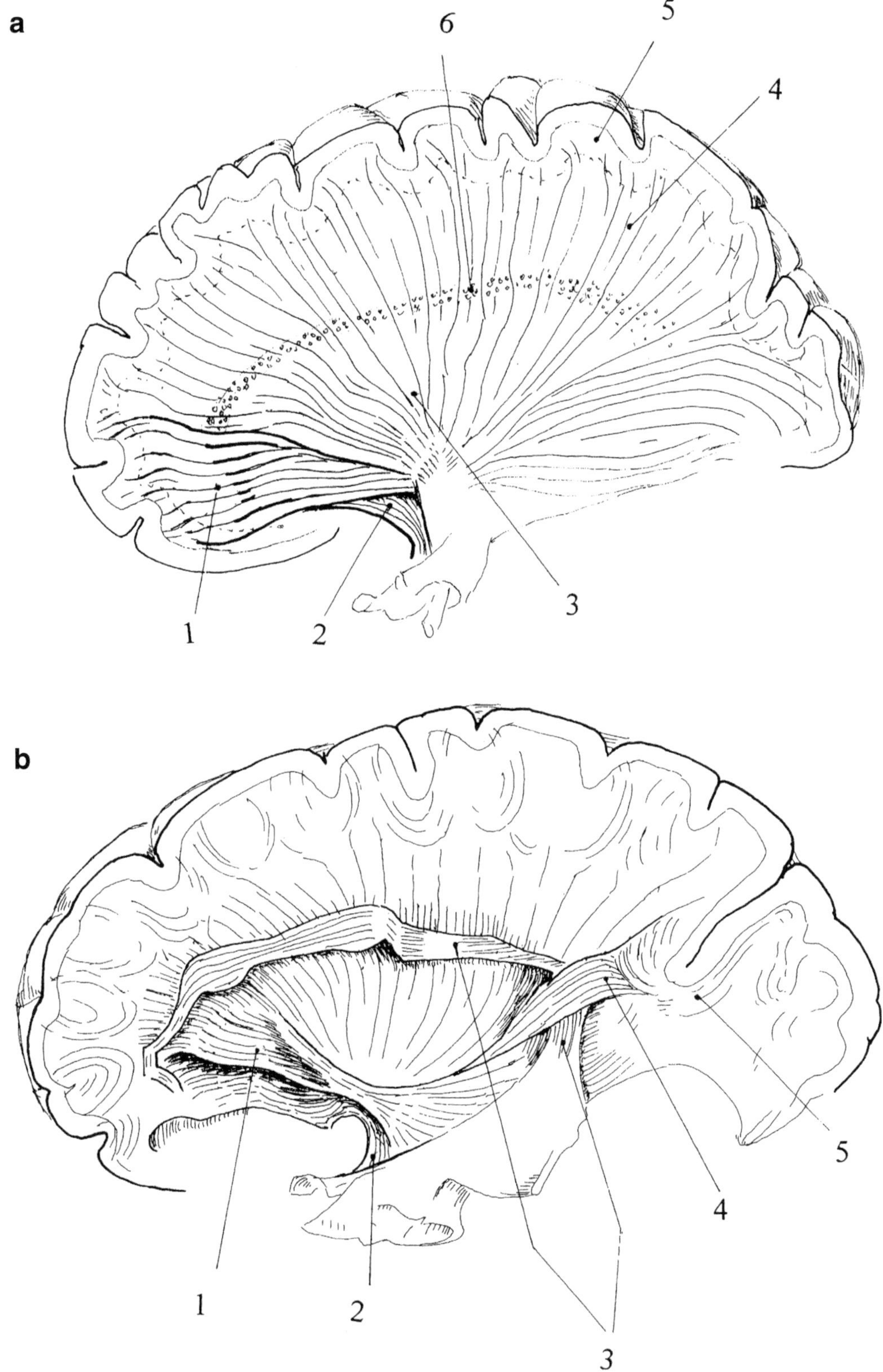

Fig. 6.8 *Association and projection fibers*. (**a**) Medial from basal ganglia. (*1*) Fasciculus occipitofrontalis, (*2*) fasciculus uncinatus, (*3*) capsula interna, (*4*) corona radiata, (*5*) U-fibers, (*6*) fibers of corpus callosum, (*4*) and (*6*) projection fibers. (**b**) Lateral from basal ganglia. (*1*) Fasciculus frontooccipitalis, (*2*) fasciculus uncinatus, (*3*) fasciculus frontoparietotemporalis, (*4*) forceps posterior of corpus callosum

6.1.3 Projection Fibers (Figs. 6.9, 6.10, 6.11, 6.12, 6.13, 6.14, 6.15, and 6.16)

6.1.3.1 Intratelencephalic Projection System (Corpus Callosum): Survey (Figs. 6.9 and 6.10)

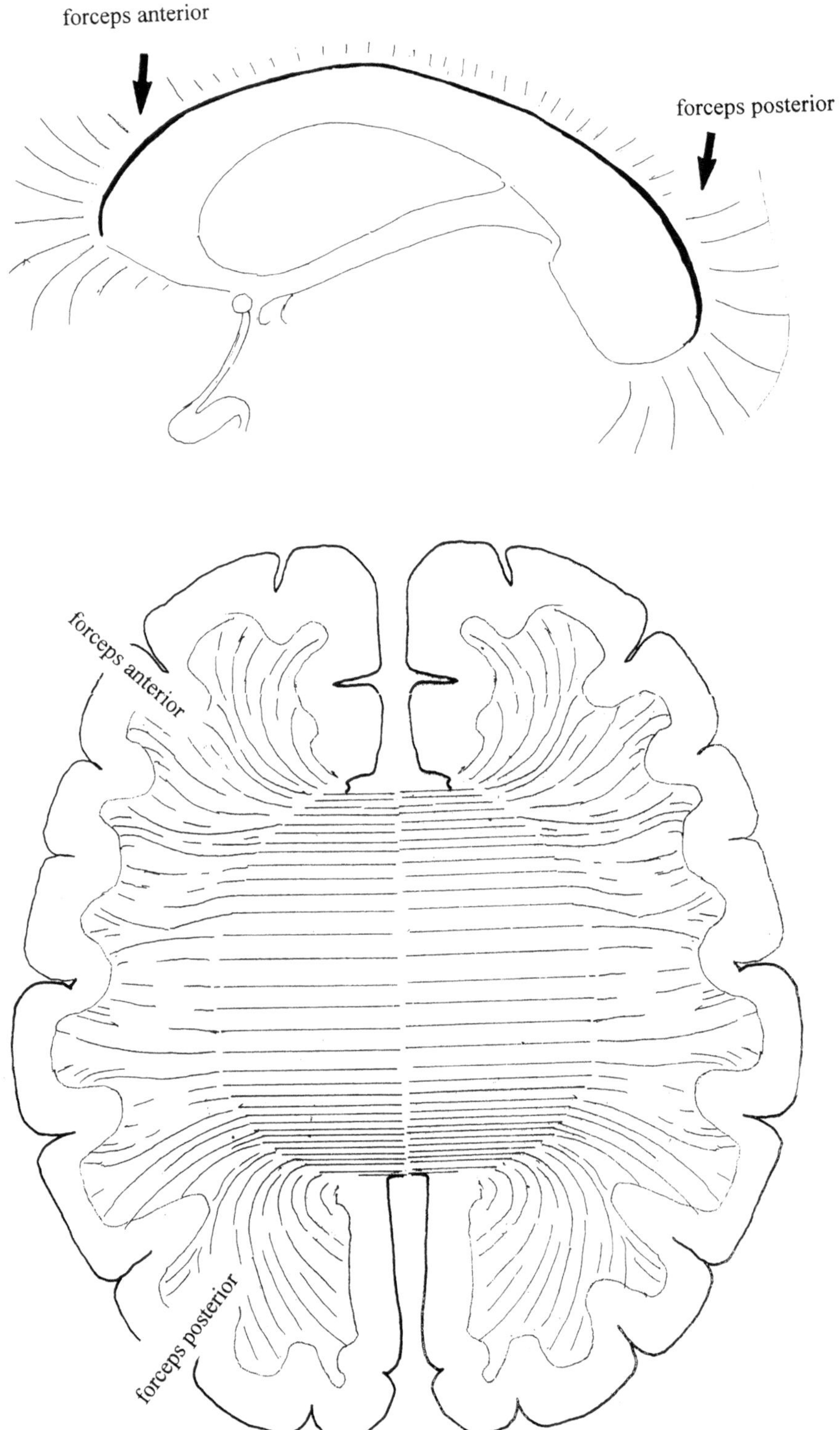

Fig. 6.9 *Projection fibers intratelencephalic (Corpus callosum). Bihemispheric exact systematic neocortical fibers (archaic neocortical fiber system).* Schematized

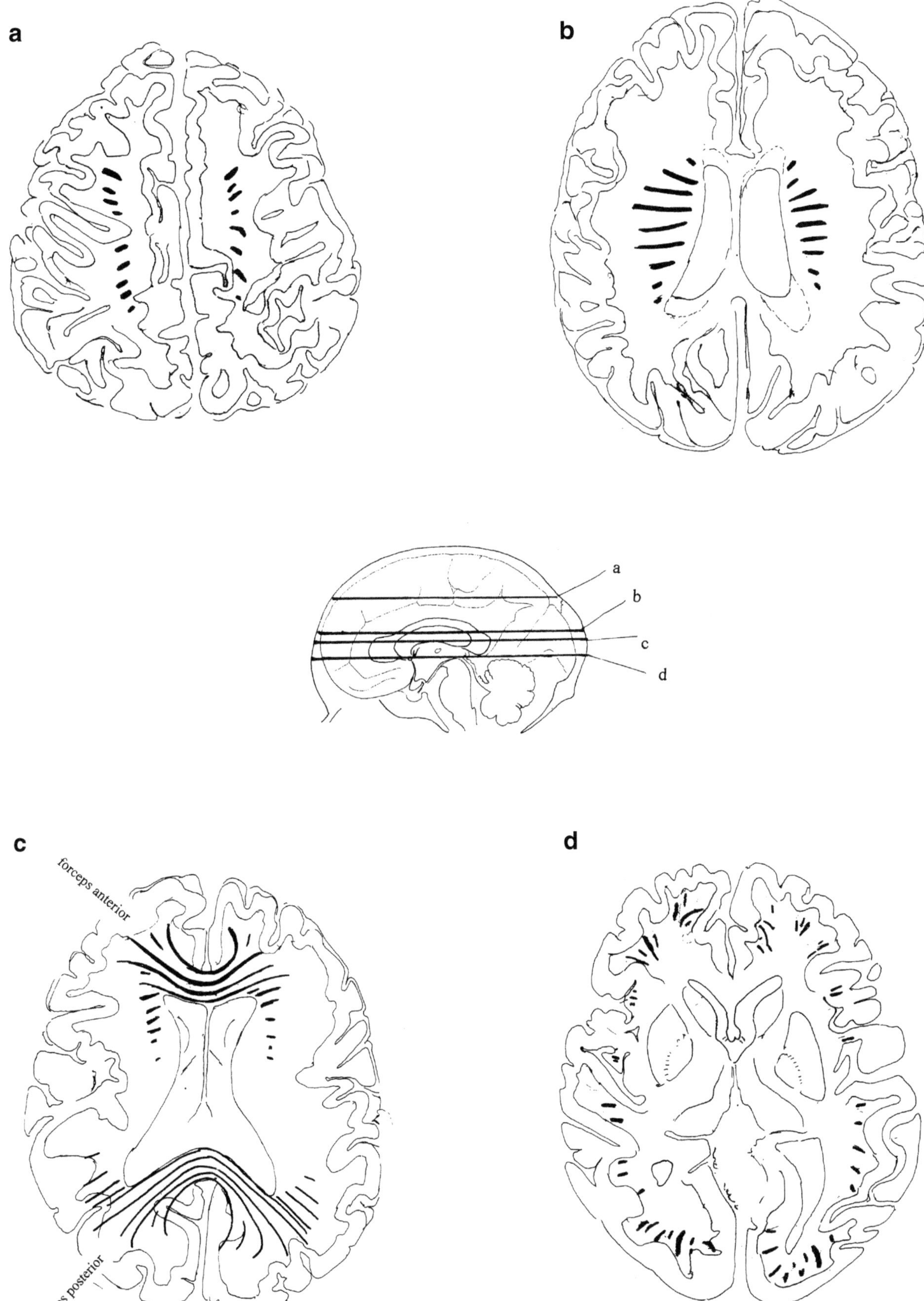

Fig. 6.10 *Continuation. MRT axial*. Fiber bundles of corpus callosum rendered prominent

6.1.3.2 Corpus Callosum and Extra-Intratelencephalic Systems (Corona Radiata) (Figs. 6.11, 6.12, 6.13, 6.14, 6.15, and 6.16)

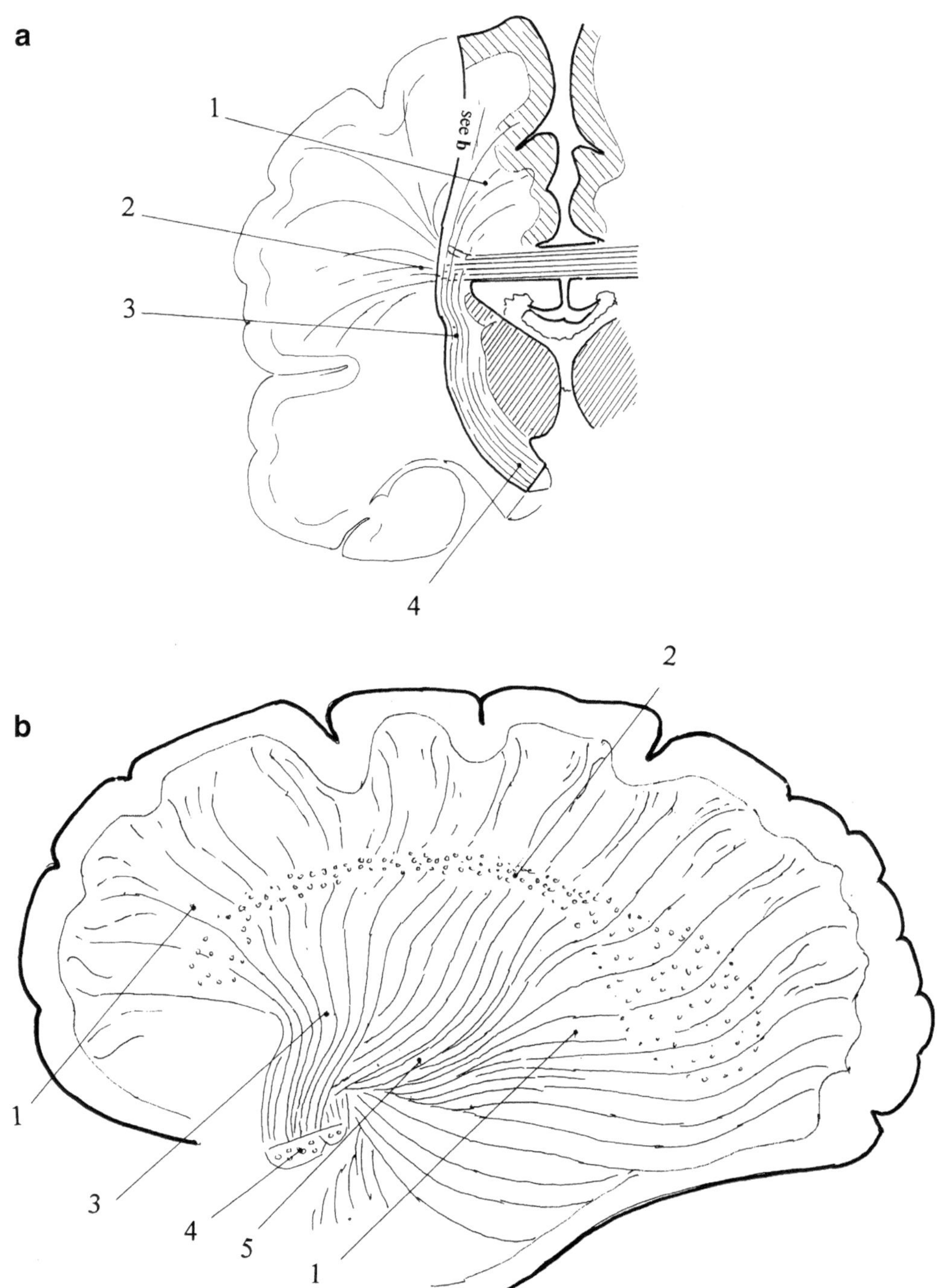

Fig. 6.11 *Fibers of corpus callosum, corona radiata: Motor (corticoefferent) and sensoric (corticoafferent) systems.* Survey. (**a**) Corona radiata, capsula interna, and crus cerebri. Topogram. (**b**) Parasagittal fiber dissection. (*1*) Corona radiata, (*2*) fiber bundles of corpus callosum, (*3*) knee of capsula interna, (*4*) crus cerebri, (*5*) crus posterius of capsula interna

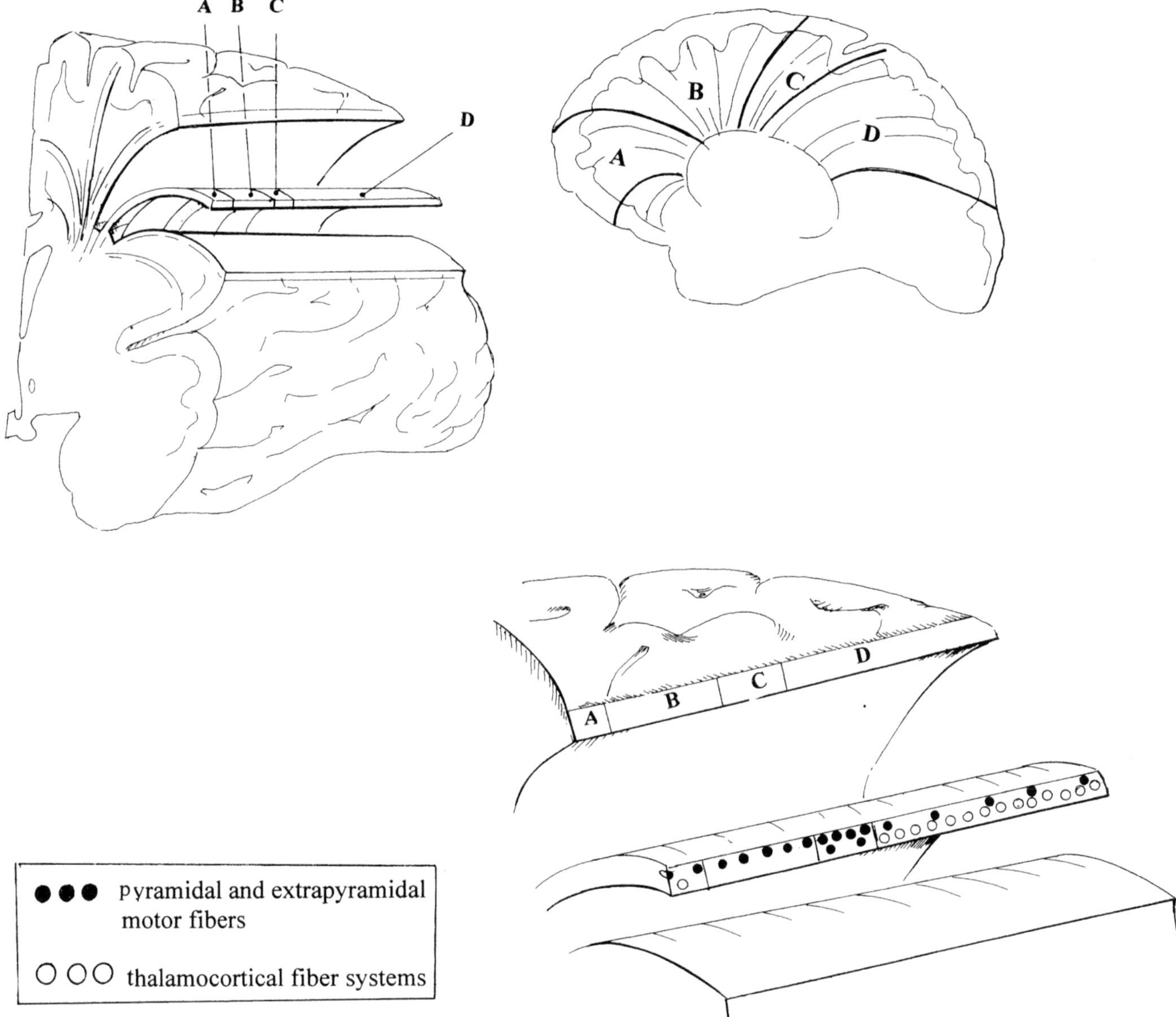

Fig. 6.12 *Architecture of corona radiata: Multiple thin-walled corticoafferent and corticoefferent fiber plates are alternating to each other.* A, B, and C are predominant corticoefferent fiber plates frontal, D are predominant corticoafferent fiber plates parietal

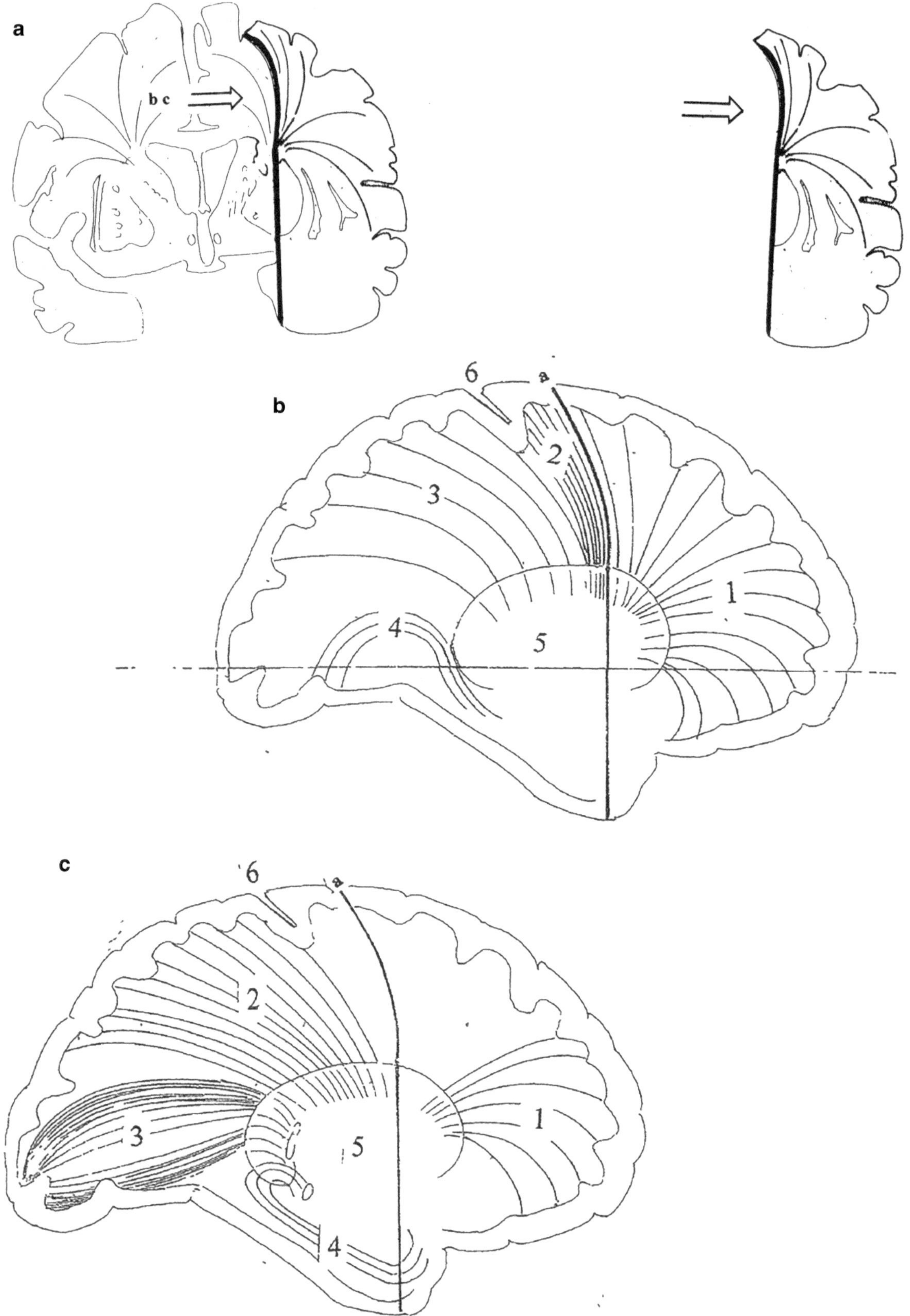

Fig. 6.13 *Details*. (**a**) Topogram for b and c, before and after elimination of the right hemisphere. Arrow: viewing direction b and c. (**b** and **c**) Motor and sensoric fibers. (*1*) Tractus frontopontinus, (*2*) tractus pyramidalis, (*3*) tractus thalamodorsalis, (*4*) acoustic fibers, (*5*) thalamus, (*6*) sulcus centralis. (**c**) Sensoric fibers. (*1*) Tractus thalamofrontalis, (*2*) tractus thalamoparietalis, (*3*) radiatio optica, (*4*) acoustic fibers, (*5*) thalamus, (*6*) sulcus centralis

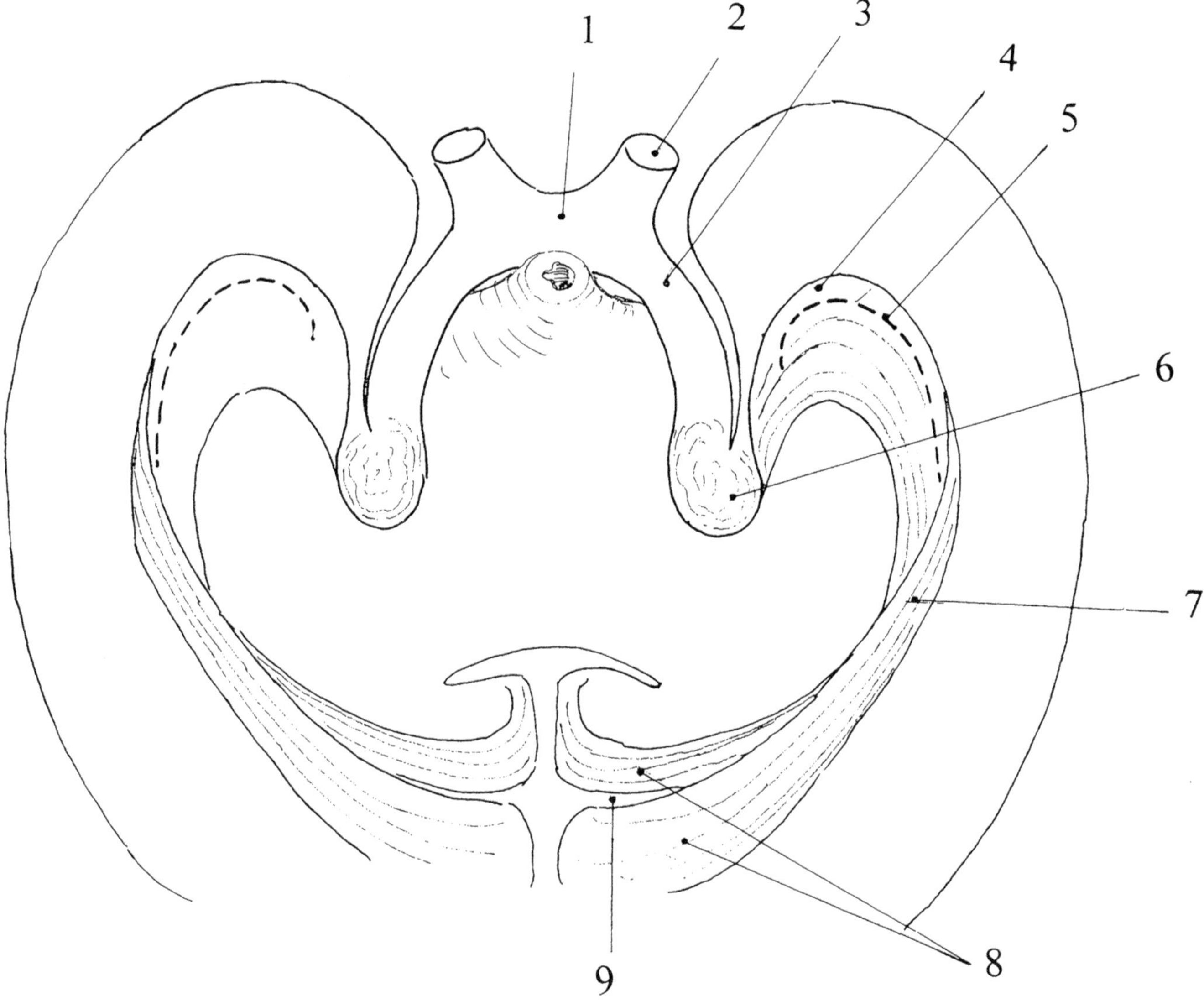

Fig. 6.14 *Radiatio optica*. Survey. (*1*) Chiasma, (*2*) n. opticus, (*3*) tractus opticus, (*4*) radiatio optica, knee, (*5*) along cornu inferius of ventriculus lateralis, (*6*) corpus geniculatum laterale, (*7*) radiatio optica at the lateral wall of cornu inferius, (*8*) cortex occipitalis, (*9*) sulcus calcarinus

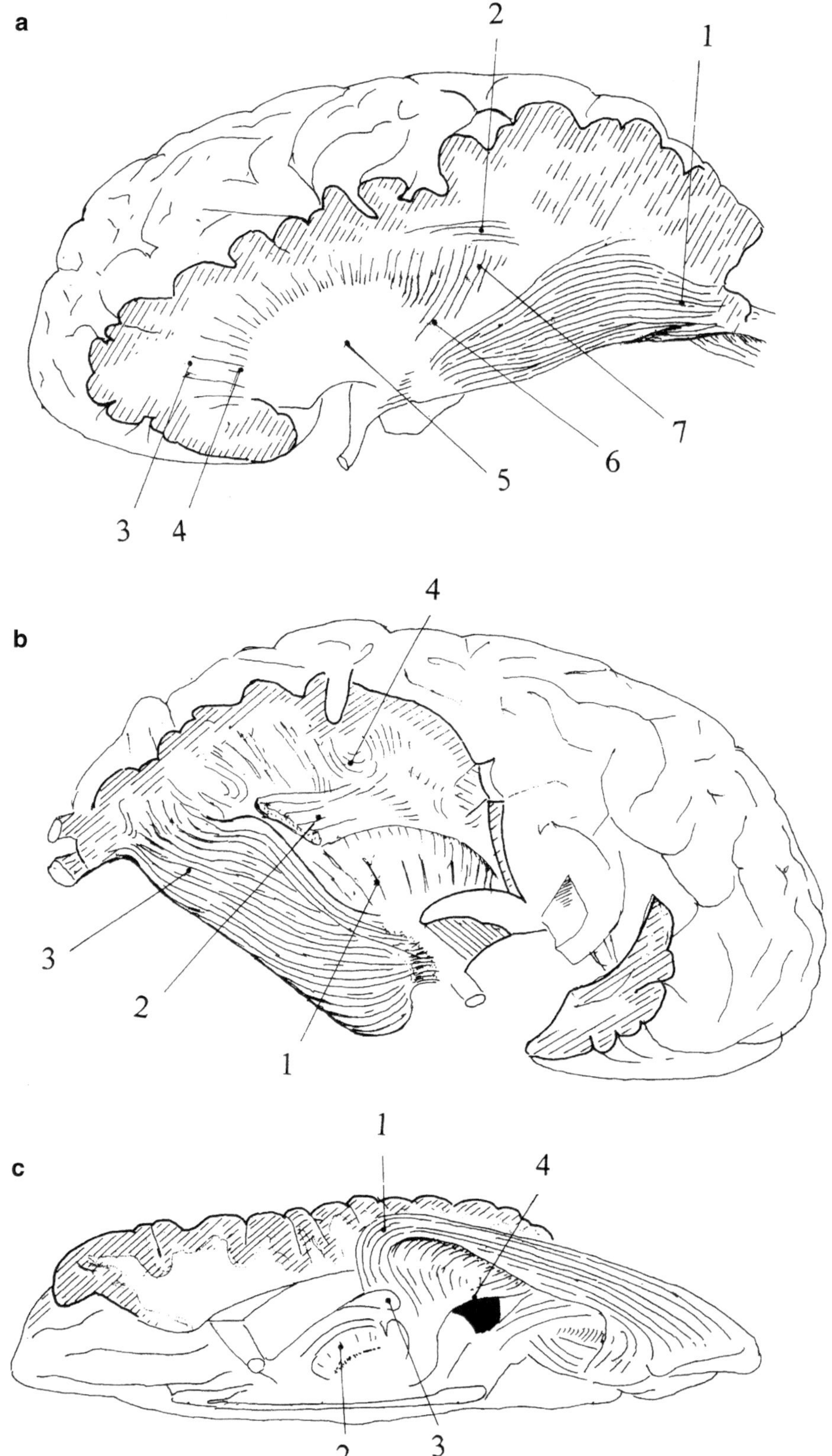

Fig. 6.15 *Radiatio optica and surrounding structures* (radiatio optica is identically presented by fiber tracking, but not its surrounding structures). *Fiber dissections* (Ludwig and Klingler [4]: Tabula 12 (a), 14 (b), 13 (c). Indian ink copies of the author, modified). (**a**) Lateral: (*1*) radiatio optica, (*2*) fasciculus longitudinalis superior, (*3*) corona radiata, (*4*) capsula interna, crus anterior, (*5*) putamen, (*6*) capsula interna, crus posterior, (*7*) as (*3*). (**b**) Medial: (*1*) corona radiata, (*2*) fasciculus longitudinalis superior, (*3*) radiatio optica, (*4*) U-fibers. (**c**) basal: (*1*) radiatio optica, genu temporale, (*2*) crus cerebri, (*3*) corpus geniculatum laterale, (*4*) atrium

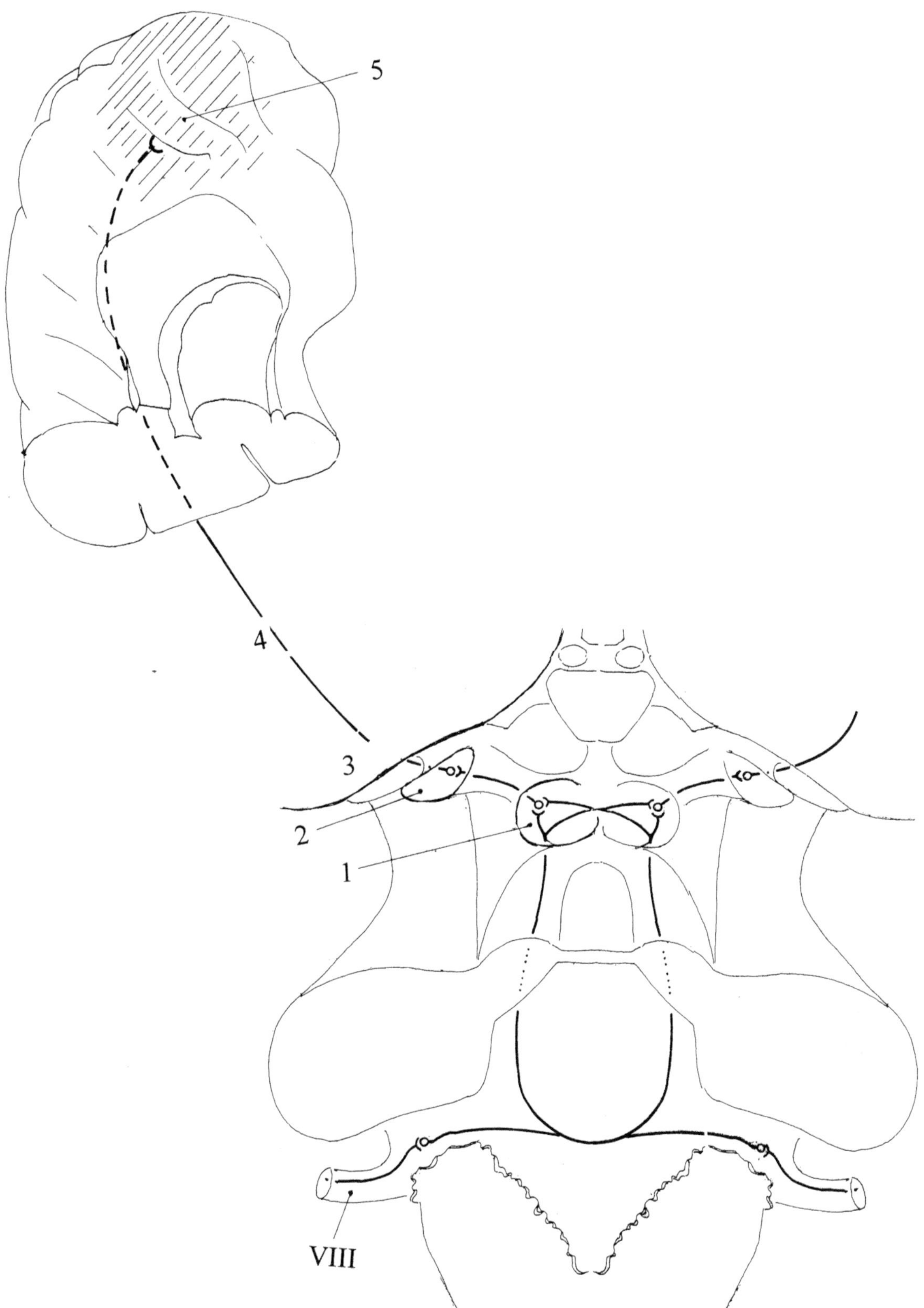

Fig. 6.16 *Acoustic fiber systems*. There are two components which are bilaterally presented with two decussationes. (VIII) n. statoacusticus and corpus trapezoideum, (*1*) lamina tecti, (*2*) corpus geniculatum laterale, (*3*) thalamus, (*4*) transthalamic temporal route, (*5*) gyri transversi (Heschl)

7 Diencephalon, Mesencephalon, and Rhombencephalon: Details

7.1 Topography (Figs. 7.1, 7.2, 7.3, and 7.4)

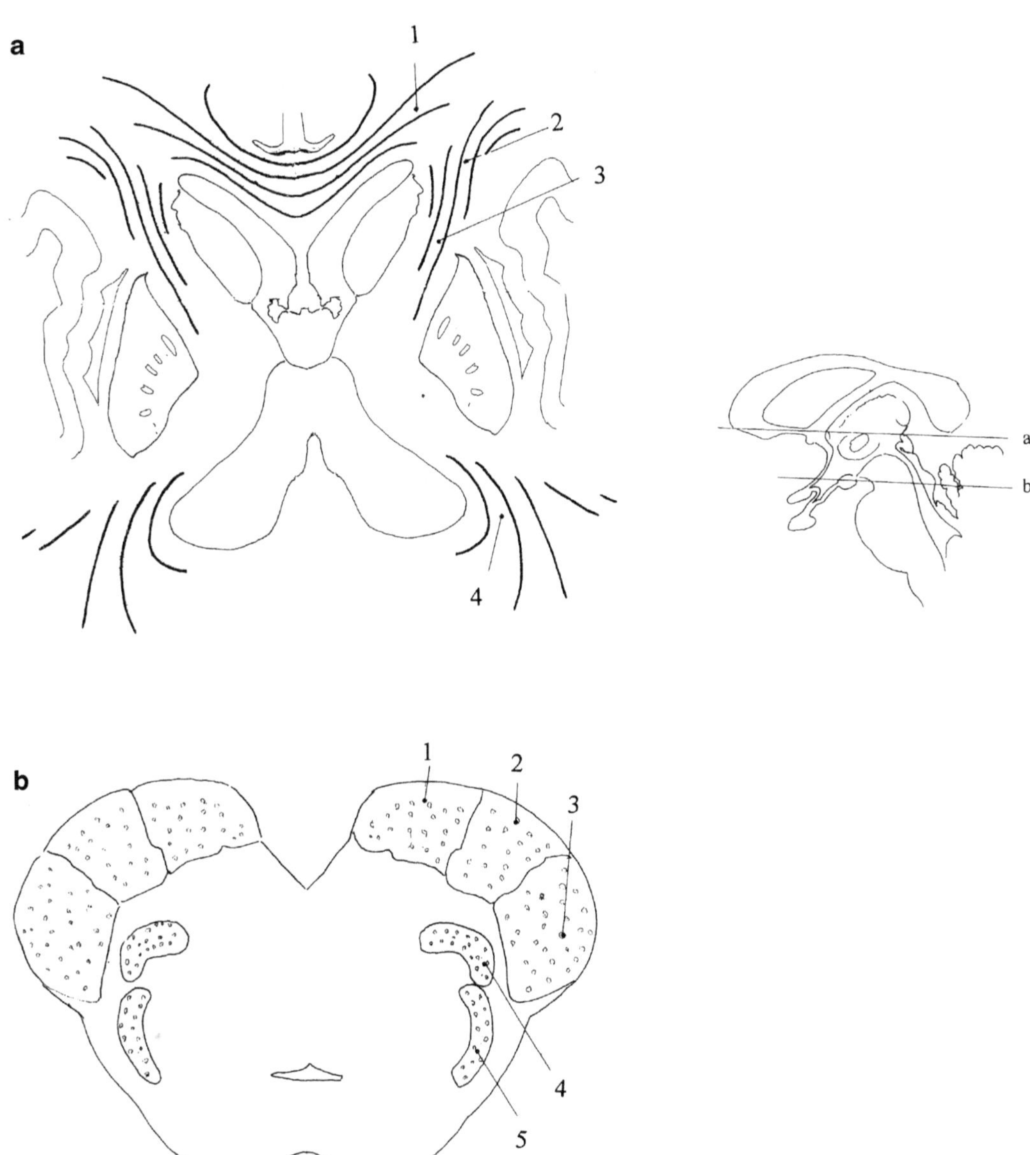

Fig. 7.1 *Topographical survey*. (**a**) Diencephalon. Axial level of foramen interventriculare, (*1*) fiber bundles of corpus callosum, (*2*) beginning of corona radiata, (*3*) crus anterius of capsula interna, (*4*) crus posterius of capsula interna. (**b**) Mesencephalon. (1) Tractus frontopontinus, (*2*) tractus pyramidalis, (*3*) tractus parietotemporo-pontinus, (*4*) lemniscus medialis, (*5*) lemniscus lateralis

W. Seeger, *Evolution of the Central Nervous System of Craniata and Homo*, https://doi.org/10.1007/978-3-030-15216-1_7

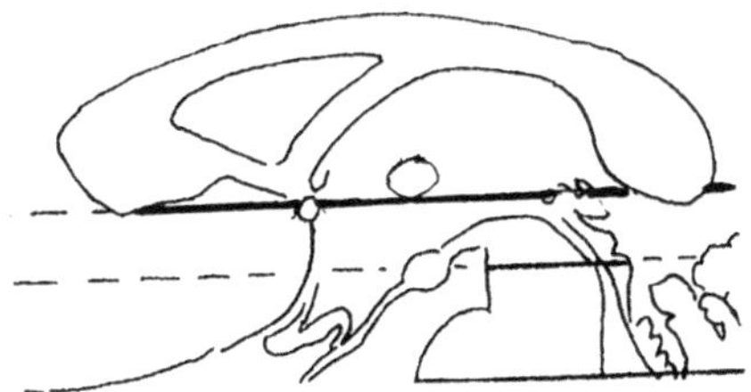

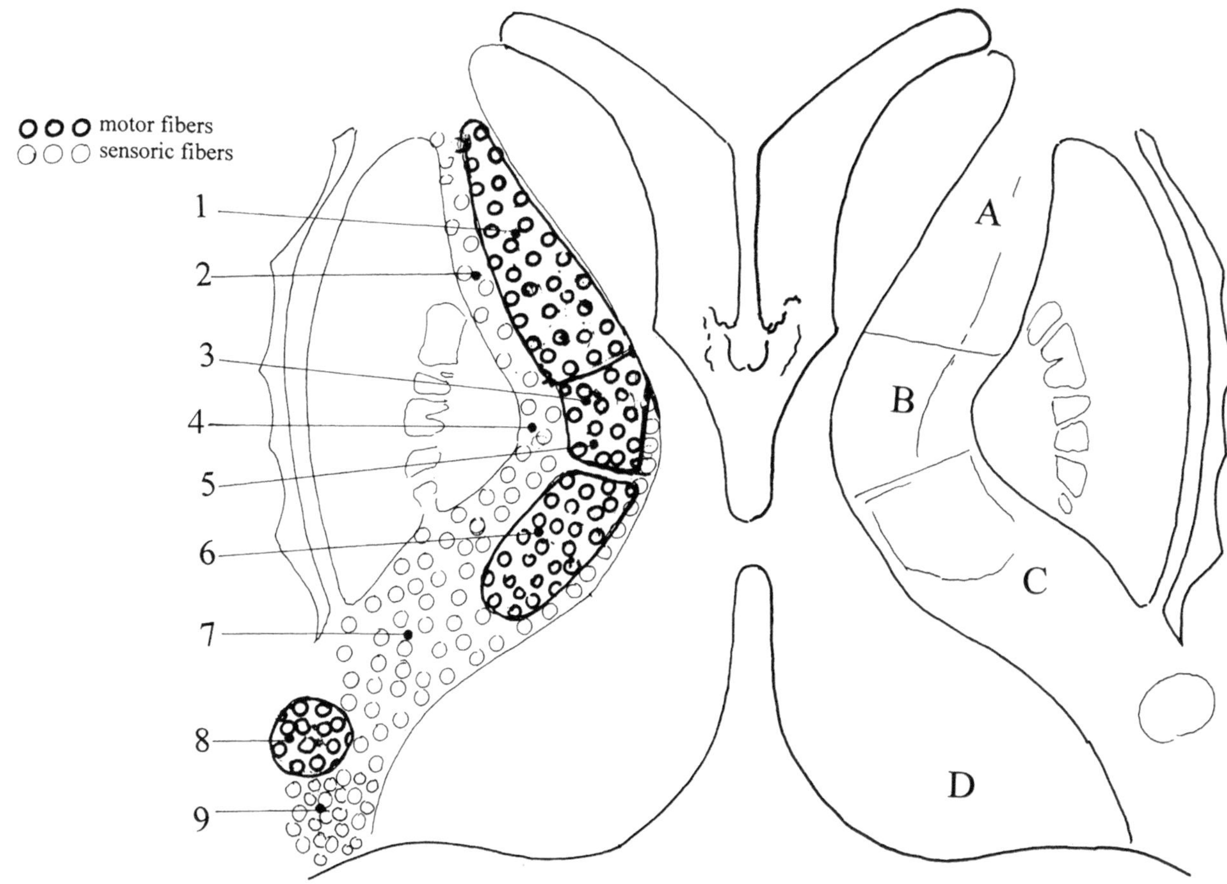

Fig. 7.2 *Addendum* (Kahle and Frotscher [3]: p. 259, modified). (A): crus anterius of capsula interna, (B) its knee, (C) crus posterius, (D) thalamus, (*1*) tractus frontopontinus, (*2*) tractus thalamofrontalis, (*3*) fibrae corticonucleares (for cranial nerves), (*4*) fibrae thalamonucleares, (*5*) as (*3*), (*6*) tractus parietotemporopontinus, (*7*) tractus thalamocorticalis parietodorsalis, (*8*) tractus temporopontinus, (*9*) radiatio optica

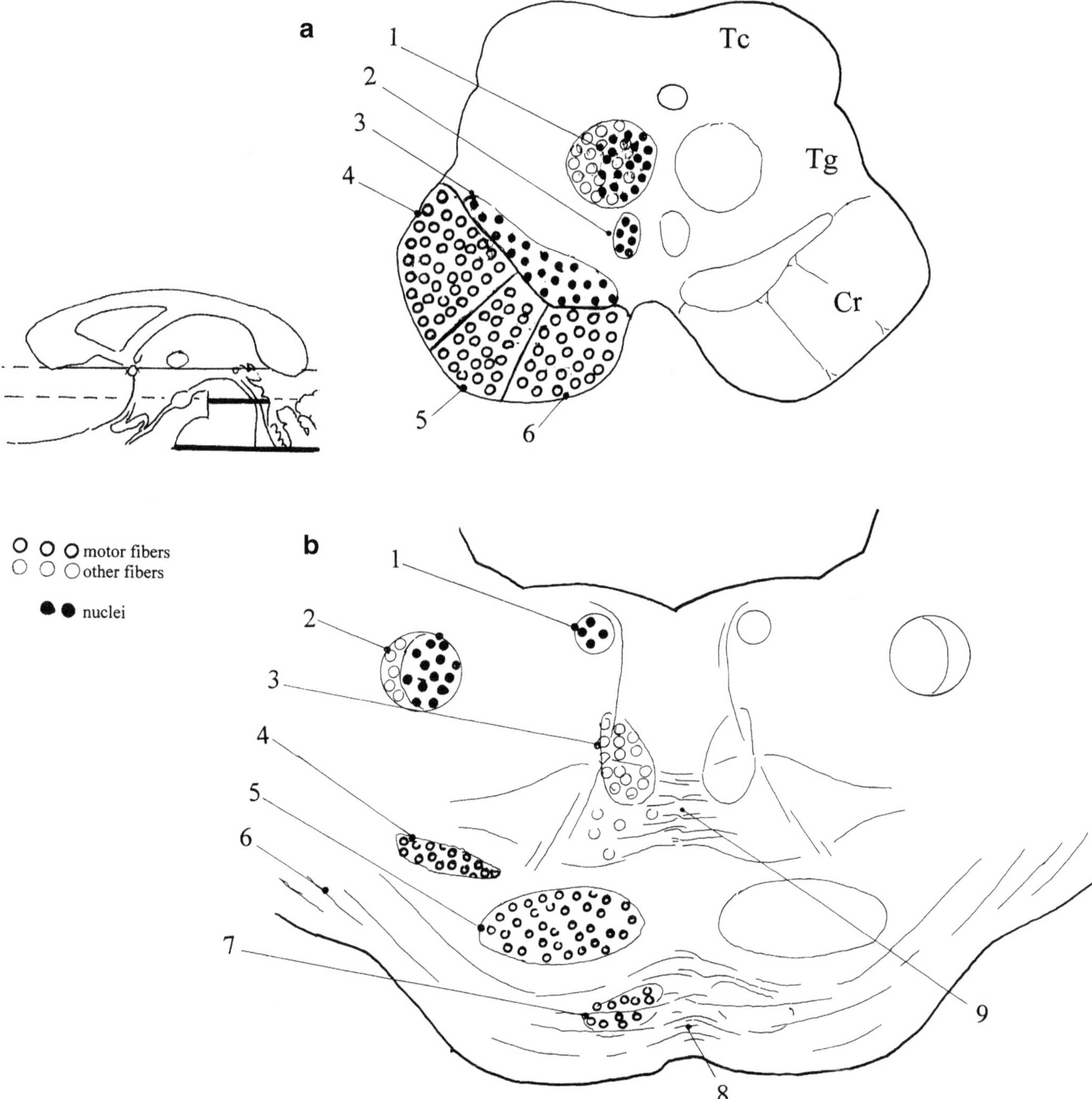

Fig. 7.3 *Mesencephalon (a) and pons (b).* **a**: *Tc* tectum, *Tg* tegmentum, *Cr* crus cerebri, (*1*) pedunculus cerebellaris superior and nucleus ruber, (*2*) nucleus nervi oculomotorii, (*3*) substantia nigra, (*4*) tractus parieto-temporopontinus, (*5*) tractus pyramidalis, (*6*) tractus frontopontinus. **b**: (*1*) nucleus nervi abducentis, (*2*) nervus et nucleus nervi trigemini, (*3*) lemniscus medialis, (*4*) substantia nigra, (*5*) tractus pyramidalis, (*6*) pedunculus cerebellaris medius (brachium pontis), (*7*) tractus fronto-pontinus, (*8*) decussatio pontis, (*9*) decussatio lemniscorum

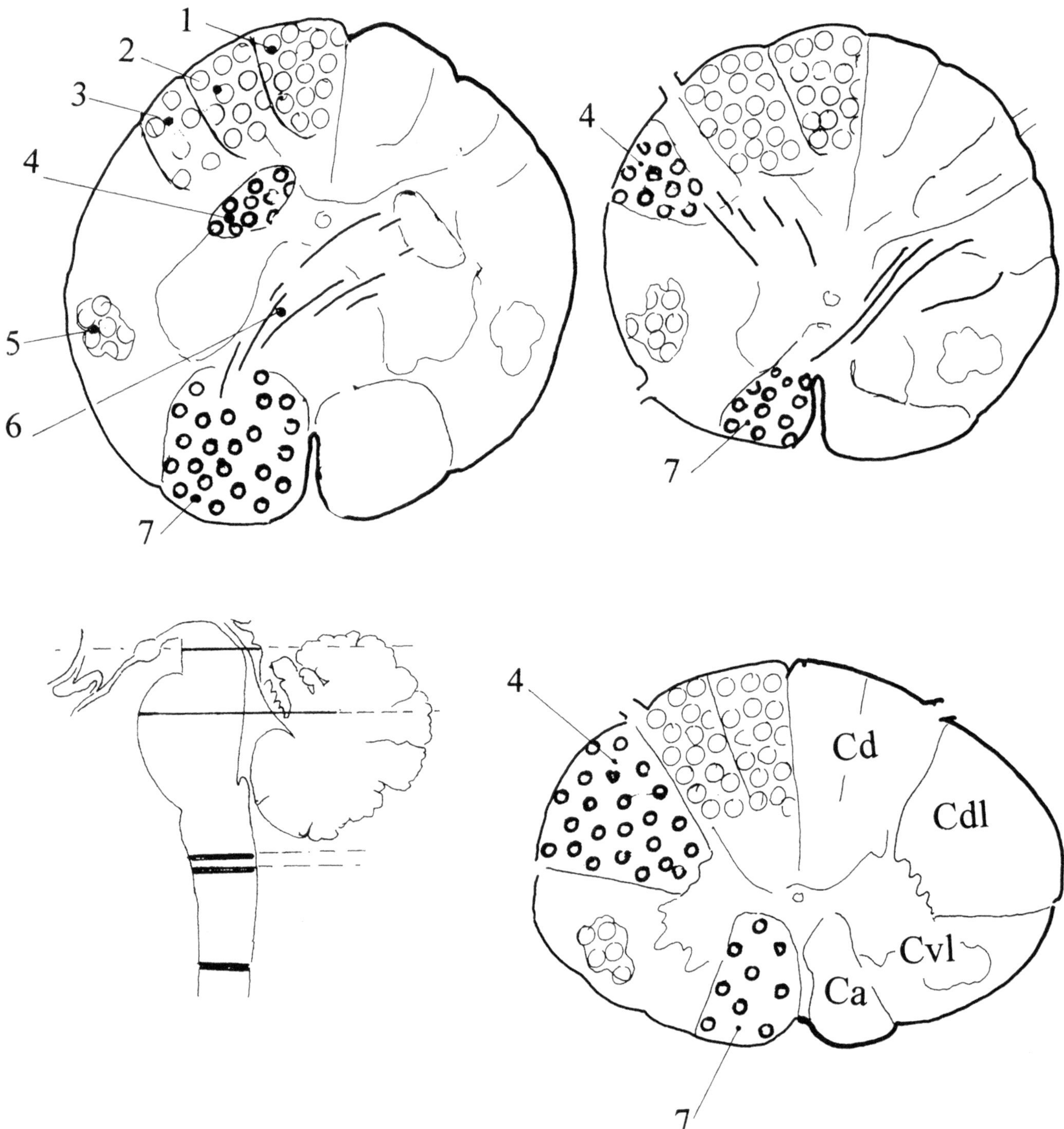

Fig. 7.4 *Medulla oblongata and medulls spinalis*. (*1*) Funiculus gracilis, (*2*) funiculus cuneatus, (*3*) tractus spinalis nervi trigemini, (*4*) tractus pyramidalis lateralis, (*5*) tractus spinothalamicus, (*6*) decussatio pyramidum, (*7*) tractus pyramidalis medialis. *Cd* columna spinalis dorsalis, *Cdl* columna dorsolateralis, *Cvl* columna ventrolateralis, *Ca* columna anterior

7.2 Fiber Connections (Figs. 7.5, 7.6, and 7.7)

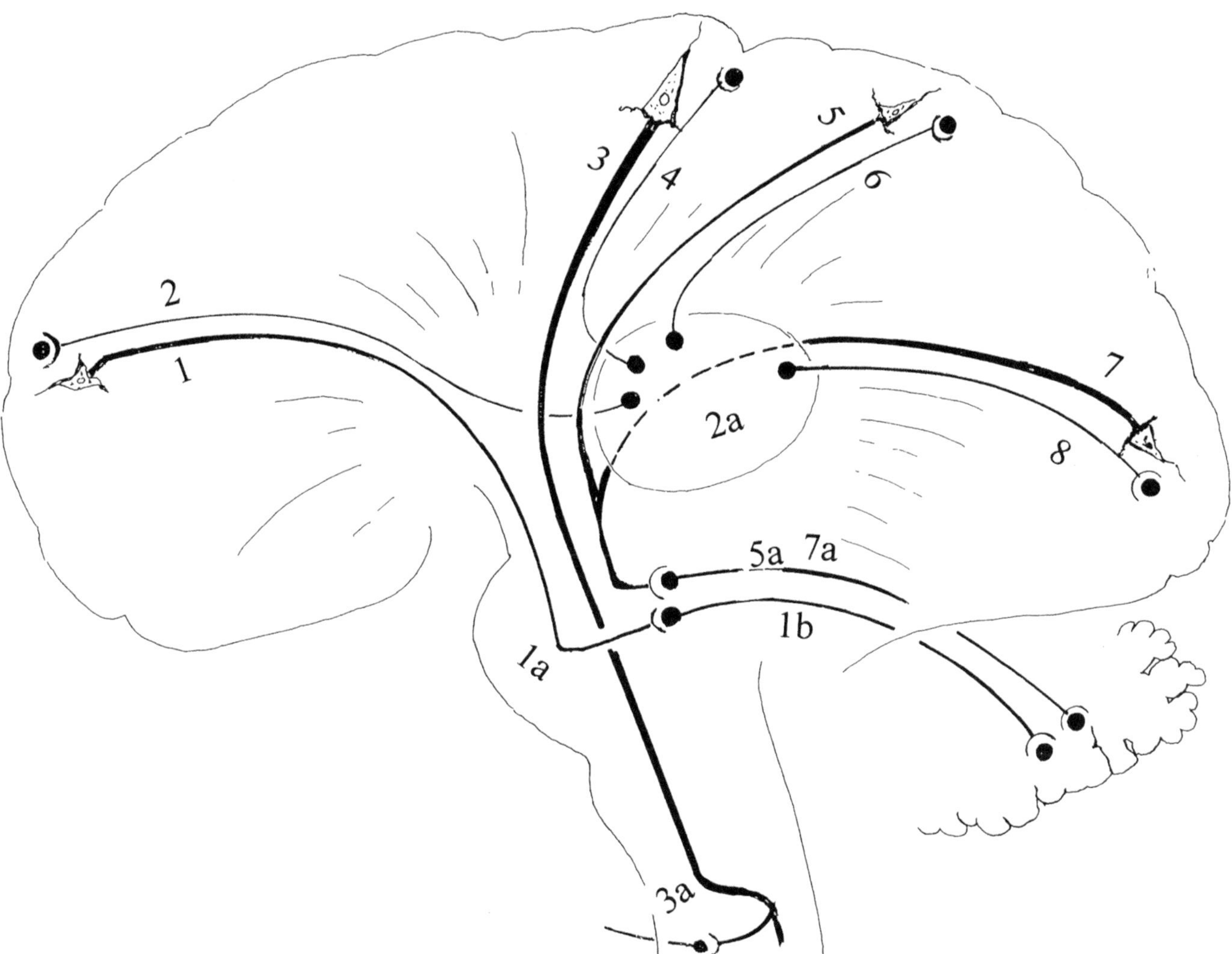

Fig. 7.5 *Motor extrapyramidal, pyramidal, and thalamocortical fiber connections*. (*1*) Tractus frontopontinus, (*1a*) decussatio pontis, (*1b*) tractus pontocerebellaris (pedunculus cerebellaris medius), (*2*) tractus thalamofrontalis, (*2a*) thalamus, (*3*) tractus pyramidalis, (*3a*) decussatio pyramidum, (*4*) tractus thalamopostcentralis, (*5*) tractus parietotemporopontinus, (*5a*) as (*7a*), (*6*) tractus thalamodorsalis, (*7*) tractus occipitopontinus, (*7a*) tractus pontocerebellaris, (*8*) radiatio optica

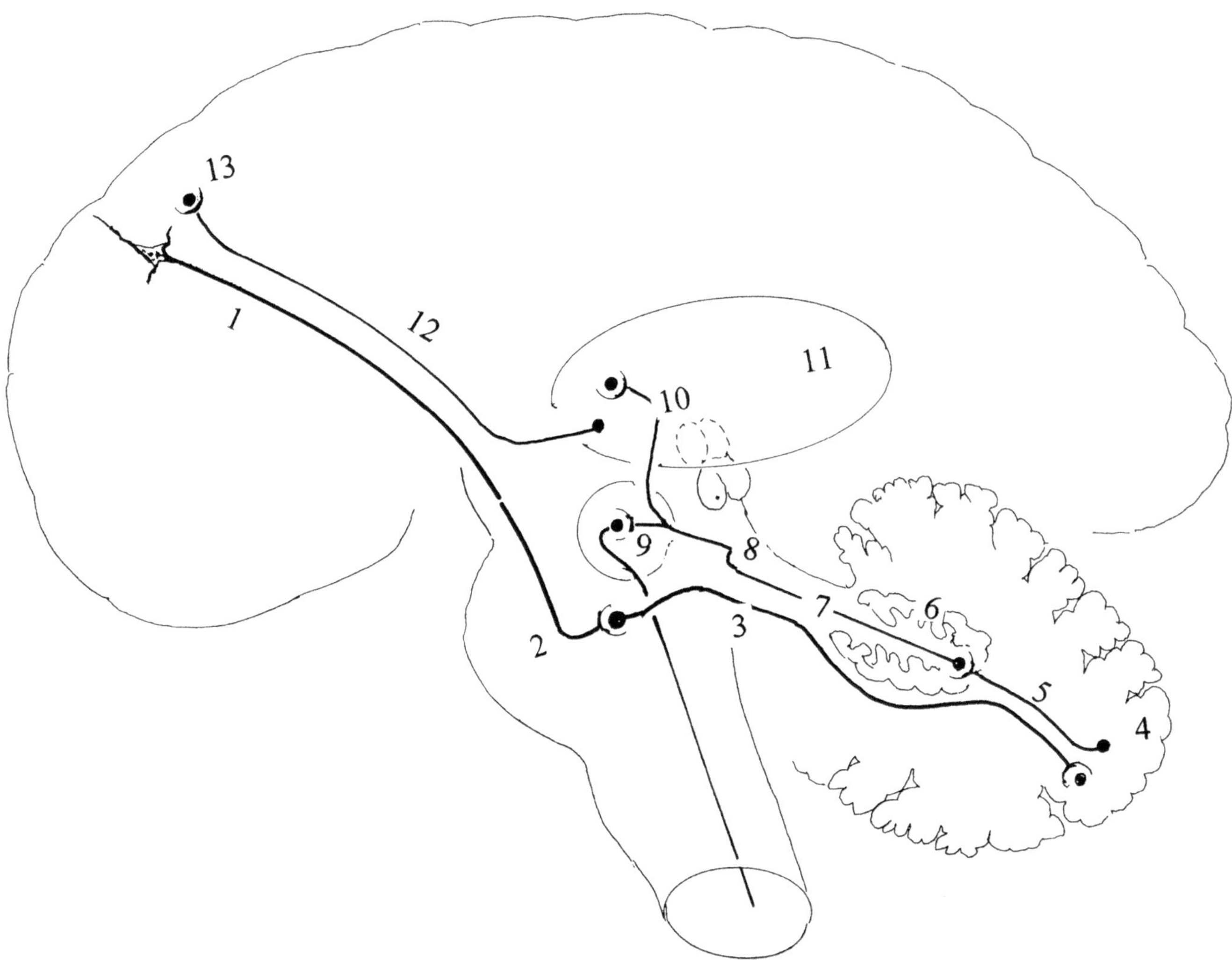

Fig. 7.6 *Addendum. Tractus frontopontinus and cerebello-mesencephalo-thalamic continuations.* (*1*) Tractus frontopontinus, (*2*) decusssatio pontis, (*3*) tractus pontocerebellaris (pedunculus pontocerebellaris medius), (*4*) cerebellar cortex, (*5*) connection of Purkinje cells and nucleus dentatus, (*6*) nucleus dentatus, (*7*) pedunculus cerebellaris superior, (*8*) decussatio of pedunculus cerebellaris superior, (*9*) nucleus ruber and tractus rubrospinalis, (*10*) connection of pedunculus cerebellaris superior with thalamus, (*11*) thalamus, (*12*) tractus thalamocorticalis frontalis, (*13*) cortex

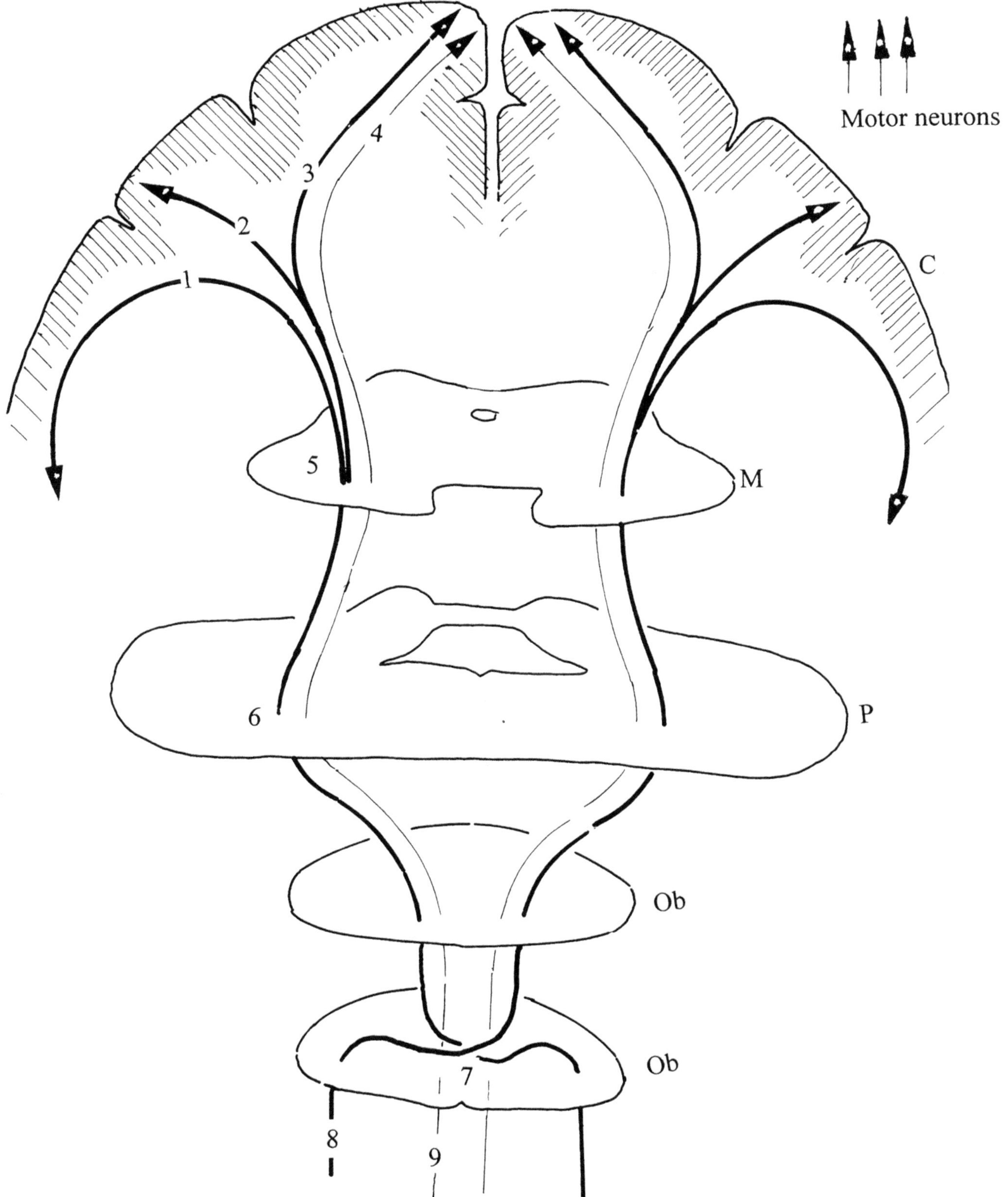

Fig. 7.7 *Tractus pyramidalis*. C gyrus praecentralis. *M* mesencephalon, *P* pons, *Ob* medulla oblongata, (*1*) fibers for cranial nerves, (*2*) fibers for hand, (*3*) for leg and bladder, (*4*) as (*3*), (*5*) crus cerebri, (*6*) pes pontis, (*7*) decussatio pyramidum, (*8*) tractus pyramidalis lateralis, 9 tractus pyramidalis medialis

8 Transectional Planes of Rhombencephalon (Medulla Oblongata, Pons) and Mesencephalon

8.1 Topography (Figs. 8.1, 8.2, 8.3, 8.4, 8.5, 8.6, and 8.7)

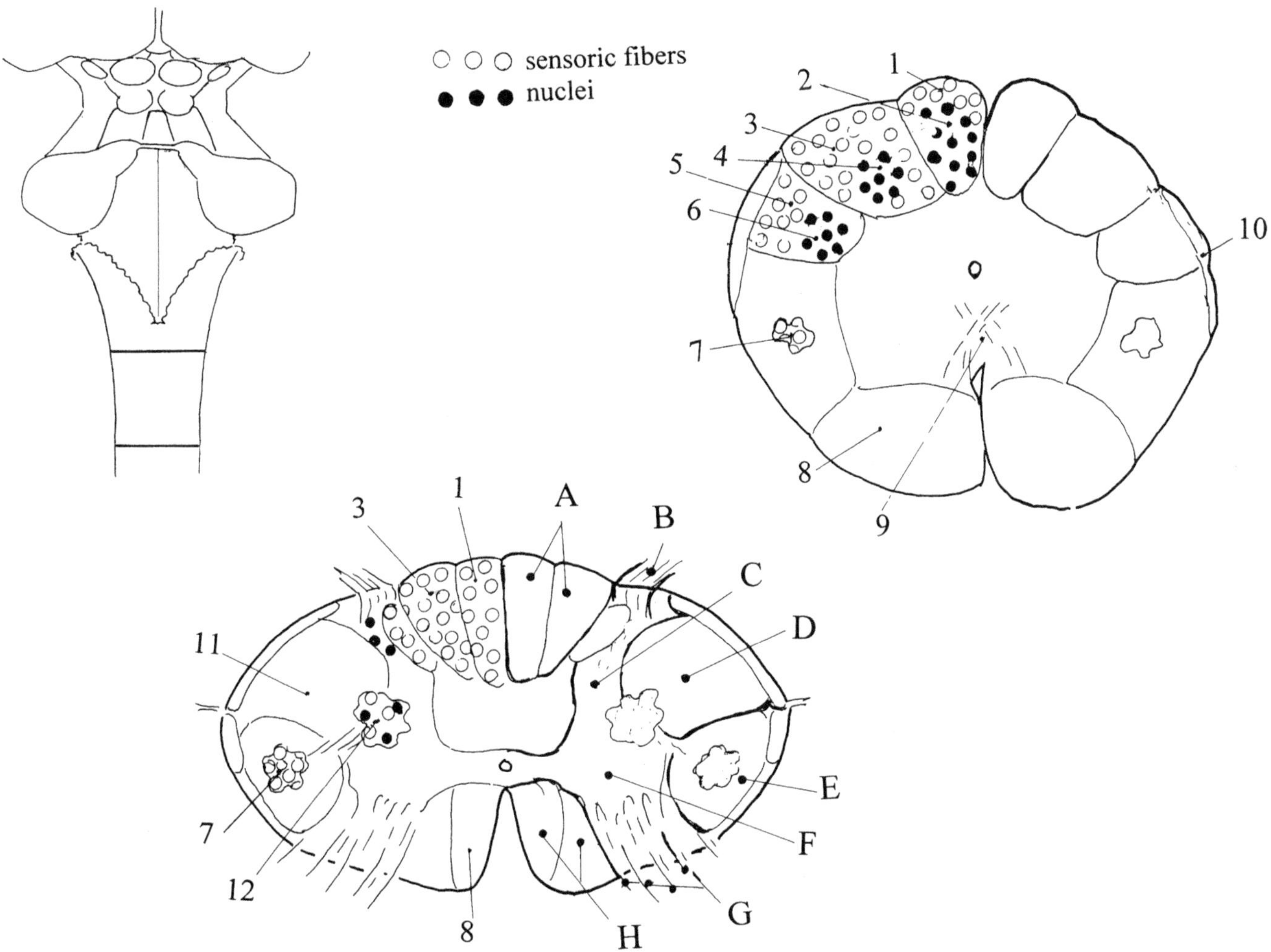

Fig. 8.1 *Medulla oblongata.* (A) Columna spinalis posterior, (B) radix spinalis posterior, (C) cornu spinalis posterius, (D) columna dorsolateralis, (E) columna ventrolateralis, (F) cornu spinale anterius, (G) radix spinalis anterior, (H) columna spinalis anterior. (*1*) Fasciculus gracilis, (*2*) its nucleus, (*3*) fasciculus cuneatus, (*4*) its nucleus, (*5*) tractus spinalis nervi trigemini, (*6*) its nucleus, (*7*) tractus spinothalamicus, (*8*) tractus spinalis anterior, (*9*) decussatio pyramidum, (*10*) tractus spinocerebellares, (*11*) tractus pyramidalis lateralis, (*12*) formatio reticularis spinalis

W. Seeger, *Evolution of the Central Nervous System of Craniata and Homo*, https://doi.org/10.1007/978-3-030-15216-1_8

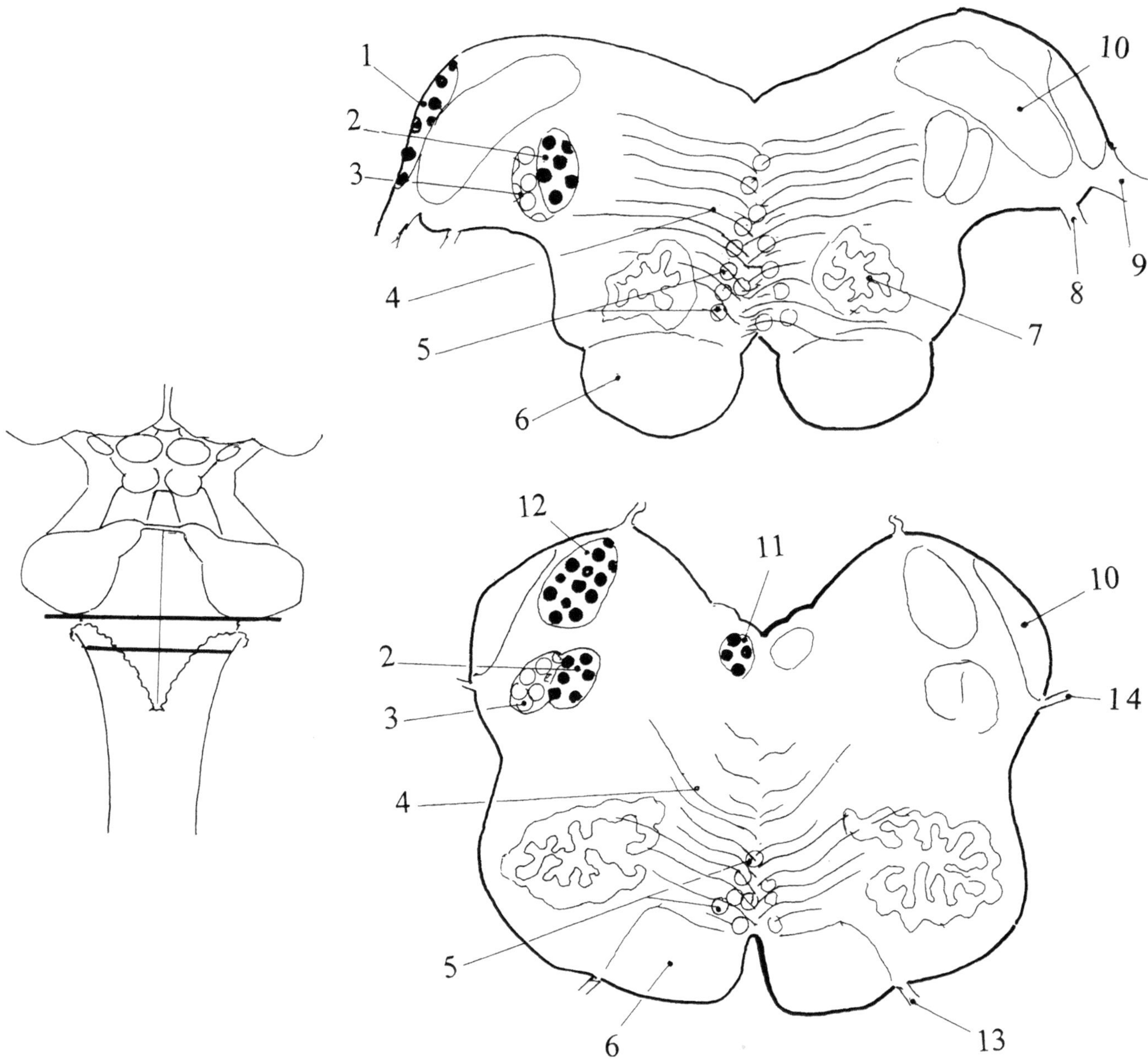

Fig. 8.2 *Superior region of medulla oblongata.* (*1*) Nucleus dorsalis nervi acustici, (*2*) nucleus spinalis nervi trigemini, (*3*) tractus spinalis nervi trigemini, (*4*) decussatio lemniscorum, (*5*) lemniscus medialis, (*6*) tractus pyramidalis, (*7*) oliva, (*8*) n. glossopharyngeus, (*9*) statoacusticus, (*10*) pedunculus cerebellaris inferior, (*11*) nucleus nervi abcucentis, (*12*) nucleus fasciculi cuneati, (*13*) n. hypoglossus, (*14*) n. accessorius

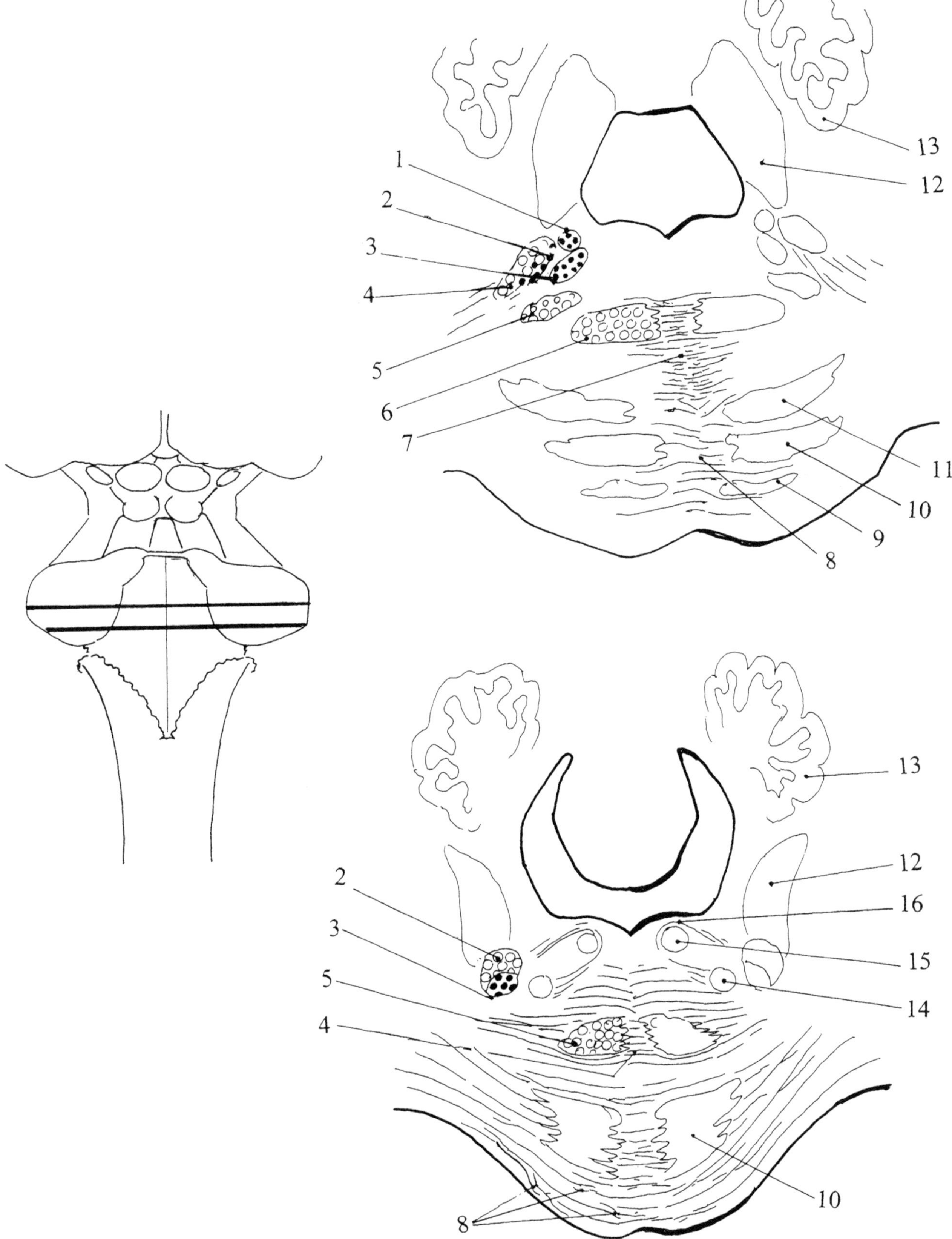

Fig. 8.3 *Inferior pontine region*. (*1*) Nucleus mesencephalicus and nucleus motorius (decussatio motorica is a part of 8, close to tegmentum pontis) nervi trigemini, (*2*) nucleus tractus spinalis nervi trigemini, (*3*) nucleus sensorius principalis nervi trigemini, (*4*) nucleus tractus spinalis nervi trigemini, (*5*) lemniscus lateralis, (*6*) lemniscus medialis, (*7*) decussatio lemniscorum, (*8*) decussatio pontis, (*9*) tractus frontopontinus, (*10*) tractus pyramidalis, (*11*) tractus parietotemporopontinus, (*12*) pedunculus cerebellaris superior, (*13*) nucleus dentatus, (*14*) nucleus nervi facialis, (*15*) nucleus nervi abducentis, (*16*) genu internum nervi facialis (Decussatio n. facialis as 38)

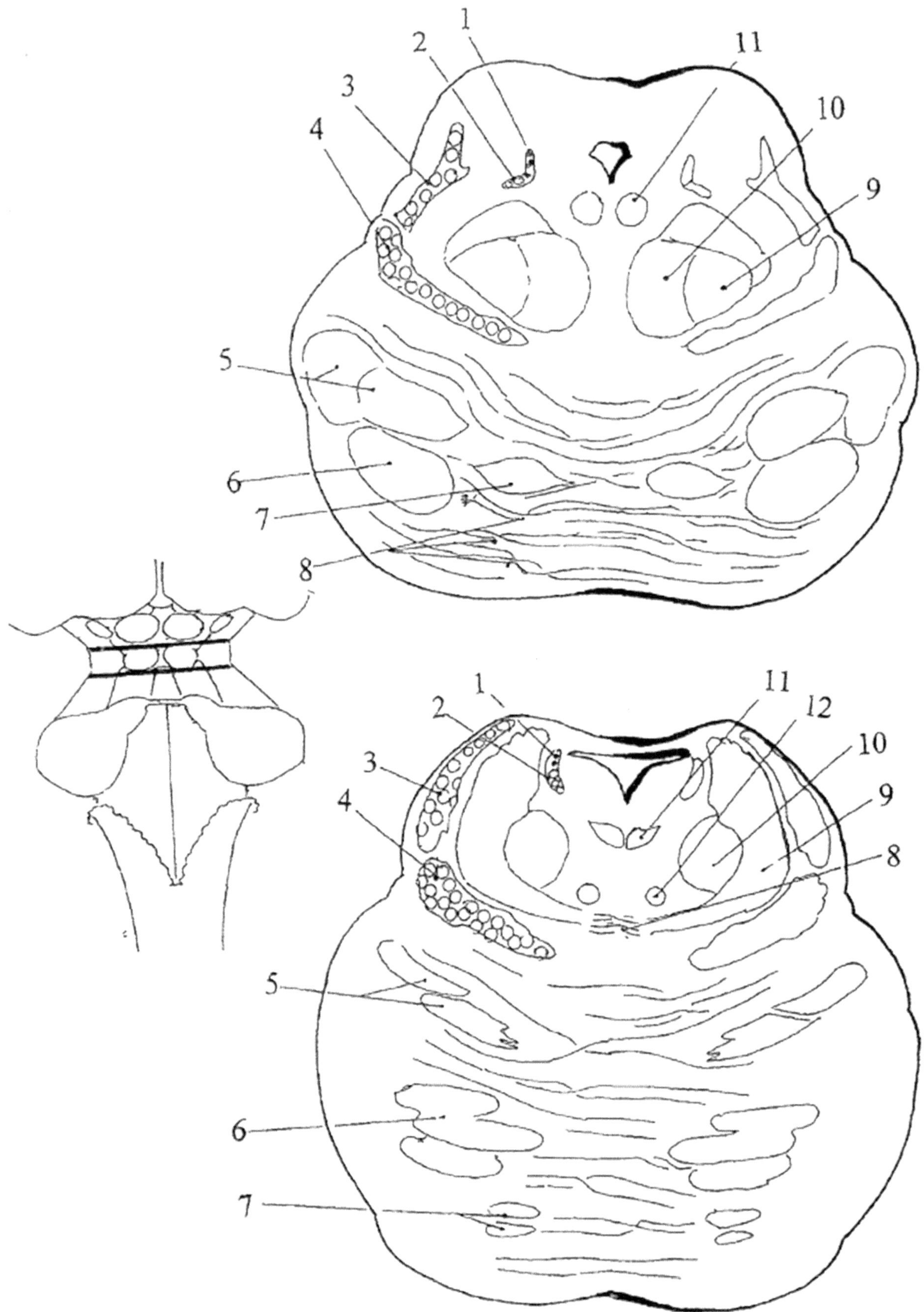

Fig. 8.4 *Pontomesencephalic levels*. (*1*) Nucleus mesencephalicus nervi trigemini, (*2*) tractus mesencephalicus nervi trigemini, (*3*) lemniscus lateralis, (*4*) lemniscus medialis, (*5*) tractus parieto-temporo-pontinus, (*6*) tractus pyramidalis, (*7*) tractus frontopontinus, (*8*) decussatio pontis, (*9*) pedunculus cerebellaris superior, (*10*) as (*9*), medial segment, (*11*) fasciculus longitudinalis medialis, (*12*) tractus rubrospinalis

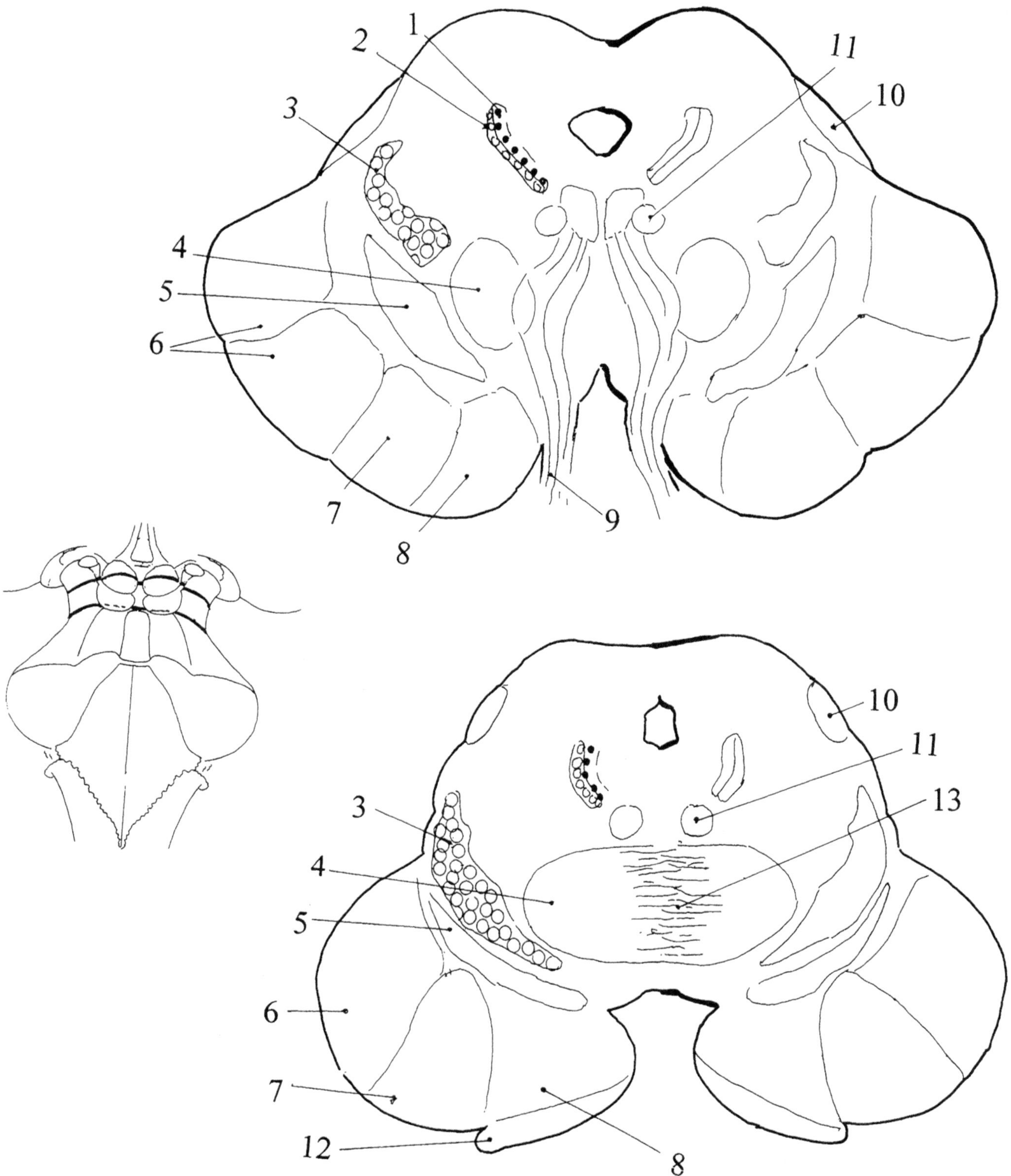

Fig. 8.5 *Intercollicular level and level of the pontomesencephalic rim.* (*1*) Nucleus mesencephalicus nervi trigemini, (*2*) tractus mesencephalicus nervi trigemini, (*3*) lemniscus medialis, (*4*) pedunculis cerebellaris superior, (*5*) substantia nigra, (*6*) tractus parietotemporopontinus, (*7*) tractus pyramidalis, (*8*) tractus frontopontinus, (*9*) n. oculomotorius, (*10*) brachium colliculi inferioris, (*11*) fasciculus longitudinalis medialis, (*12*) pons, (*13*) decussatio of pedunculus cerebellaris superior

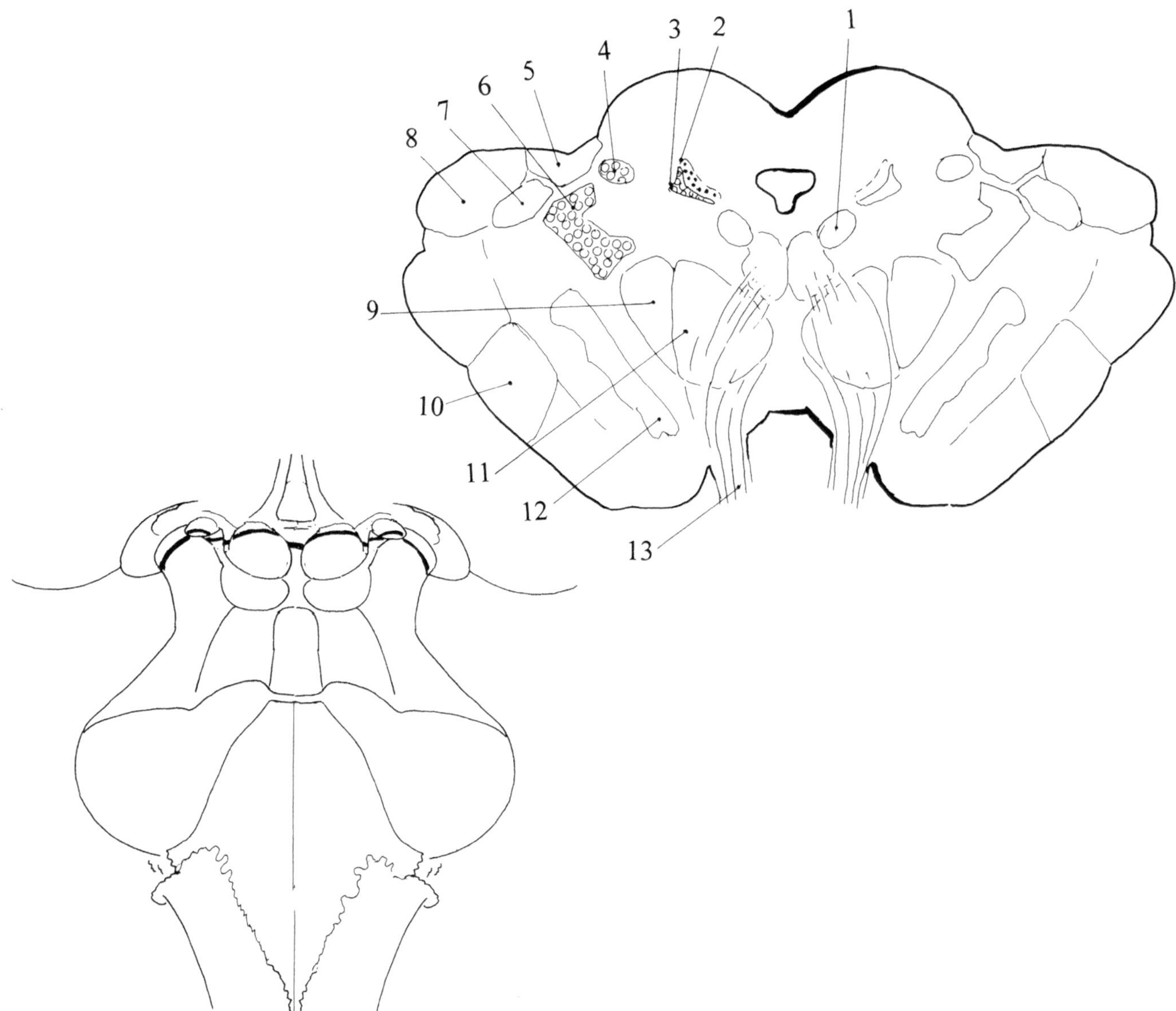

Fig. 8.6 *Level of corpus geniculatum mediale.* (*1*) Fasciculus longitudinalis medialis, (*2*) nucleus mesencephalicus nervi trigemini, (*3*) tractus mesencephalicus nervi trigemini, (*4*) tractus spinothalamicus, (*5*) brachium colliculi superioris, (*6*) lemniscus lateralis et medialis, (*7*) brachium colliculi inferioris, (*8*) corpus geniculatum mediale, (*9*) pedunculus cerebellaris superior, (*10*) tractus pyramidalis, (*11*) nucleus ruber, (*12*) substantia nigra, (*13*) n. oculomotorius

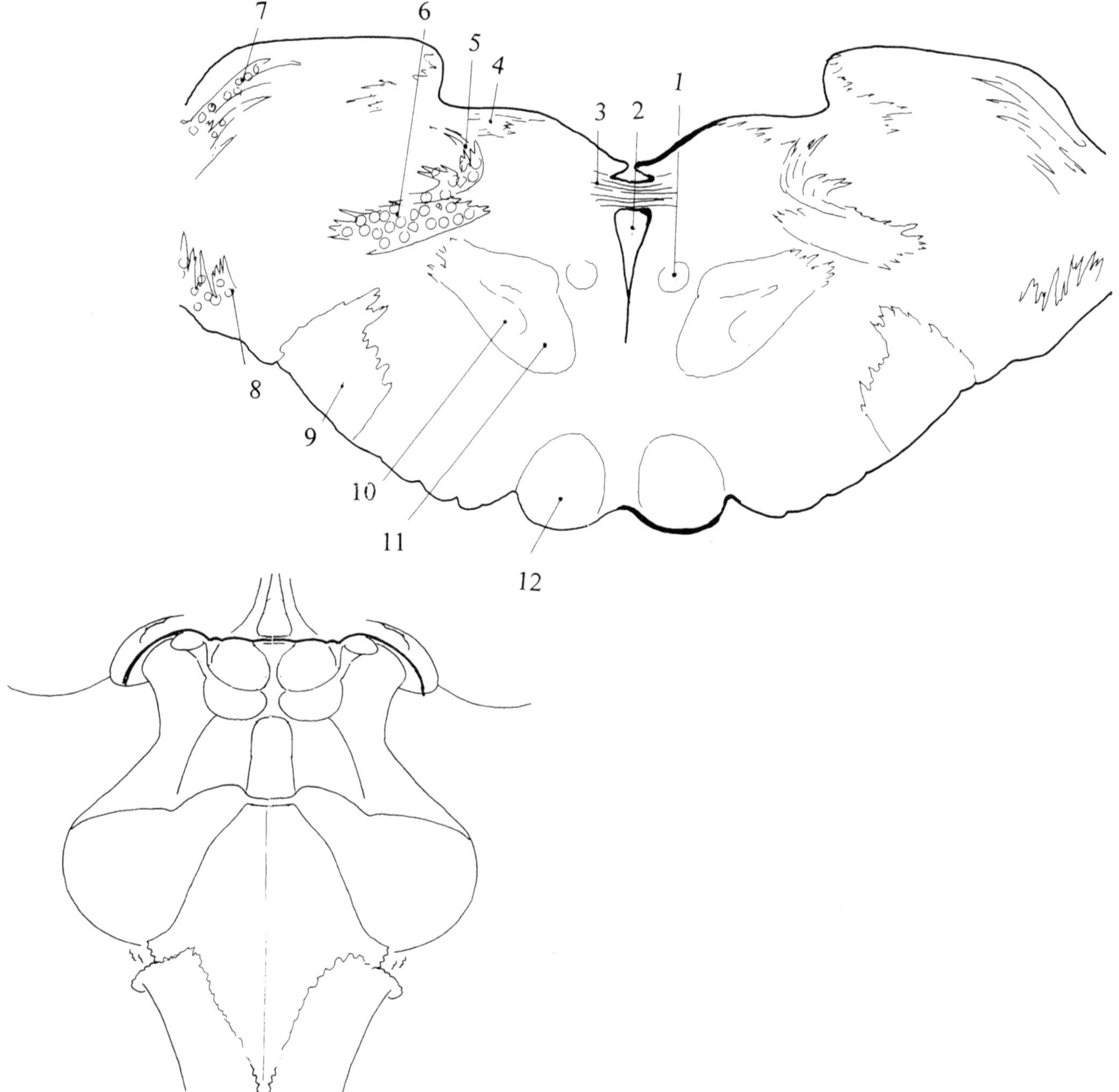

Fig. 8.7 *Level of commissura posterior*. (*1*) Fasciculus longitudinalis posterior, (*2*) third ventricle, (*3*) commissura posterior, (*4*) brachium colliculi superioris, (*5*) tractus spinothalamicus, (*6*) lemniscus medialis, (*7*) radiatio optica, (*8*) tractus opticus, (*9*) tractus pyramidalis, (*10*) pedunculus cerebellaris superior, (*11*) nucleus ruber, (*12*) corpus mamillare

8.2 Fiber Connections (Figs. 8.8, 8.9, 8.10, and 8.11)

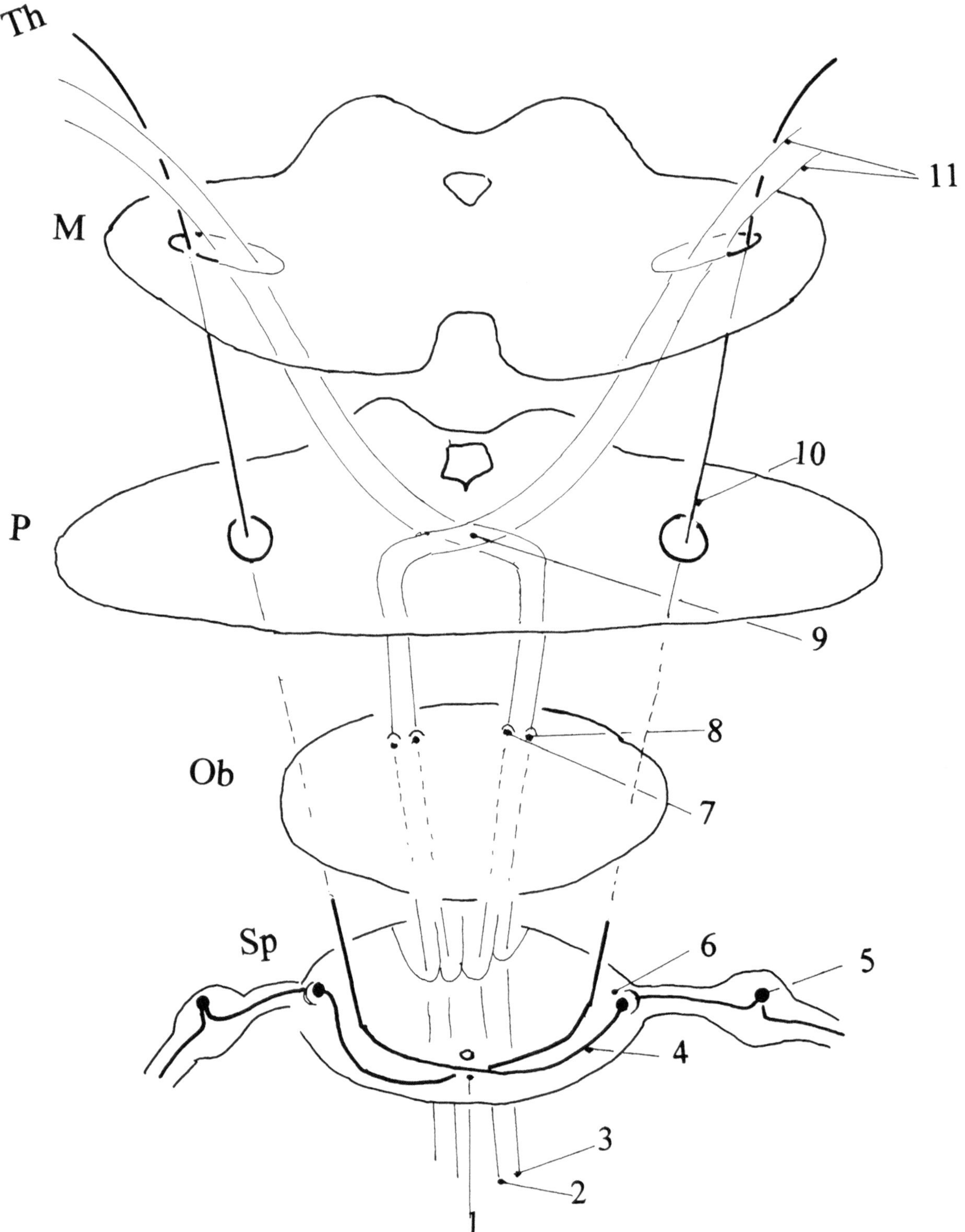

Fig. 8.8 *Tractus spinothalamicus, decussatio lemniscorum, fasciculus gracilis et cuneatus. Th* thalamus, *M* mesencephalon, *P* pons, *Ob* medulla oblongata, *Sp* medulla spinalis. (*1*) Decussatio of tractus spinothalamicus at each spinal segment, (*2*) fasciculus gracilis, (*3*) fasciculus cuneatus, (*4*) tractus spinothalamicus, (*5*) cells of ganglion spinale, (*6*) cells of cornu posterius, (*7*) nucleus fasciculi gracilis, (*8*) nucleus fasciculi cuneati, (*9*) decussatio lemniscorum, (*10*) as (*4*), (*11*) lemniscus medialis

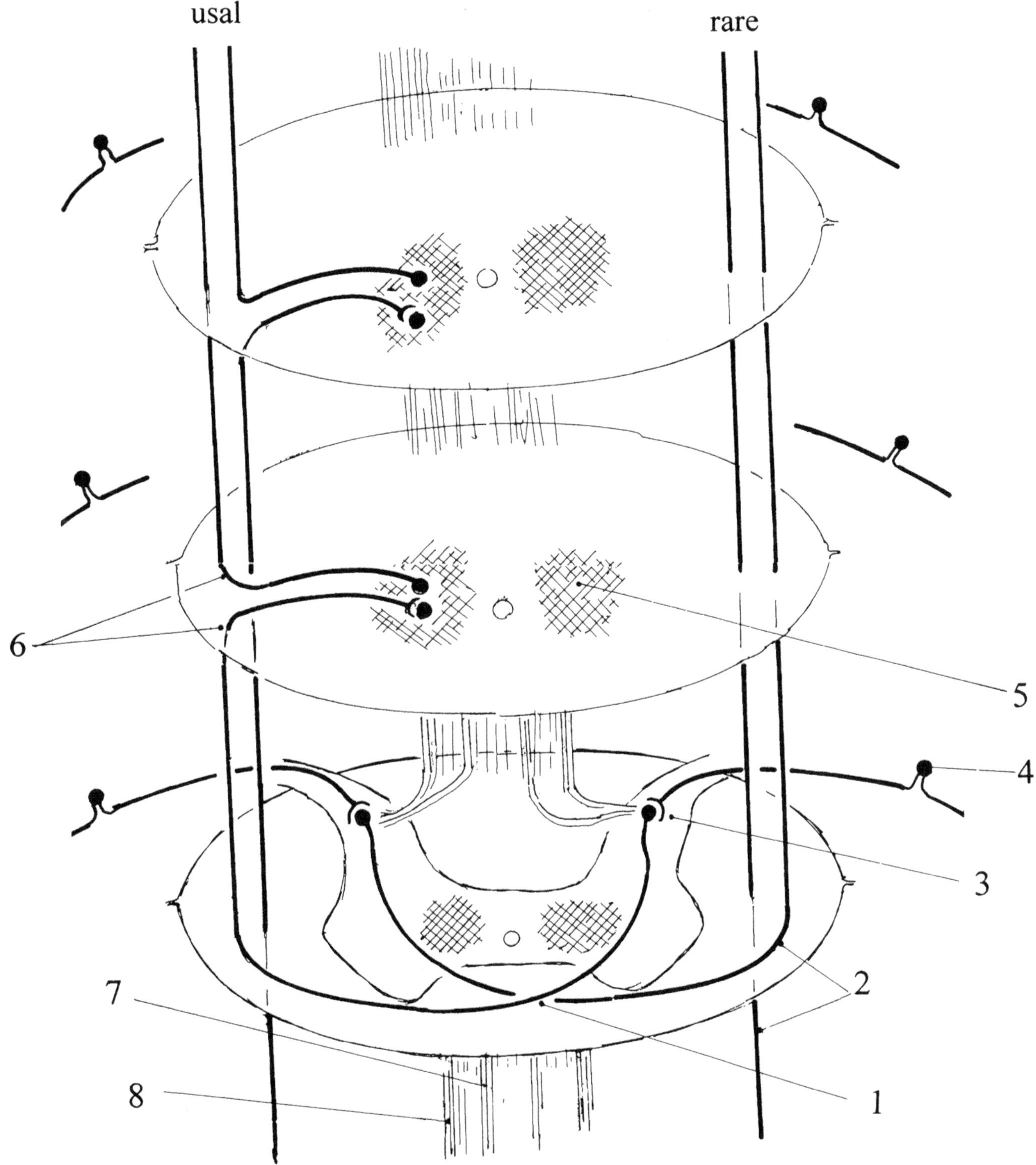

Fig. 8.9 *Tractus spinothalamicus*: Less systematic fiber system. (*1*) Decussatio of tractus spinothalamicus, (*2*) tractus spinothalamicus, (*3*) connection with each spinal segment, (*4*) spinal ganglion, (*5*) formatio reticularis spinalis, (*6*) roundabout fiber connections of tractus spinothalamicus, (*7*) fasciculus gracilis, (*8*) fasciculus cuneatus

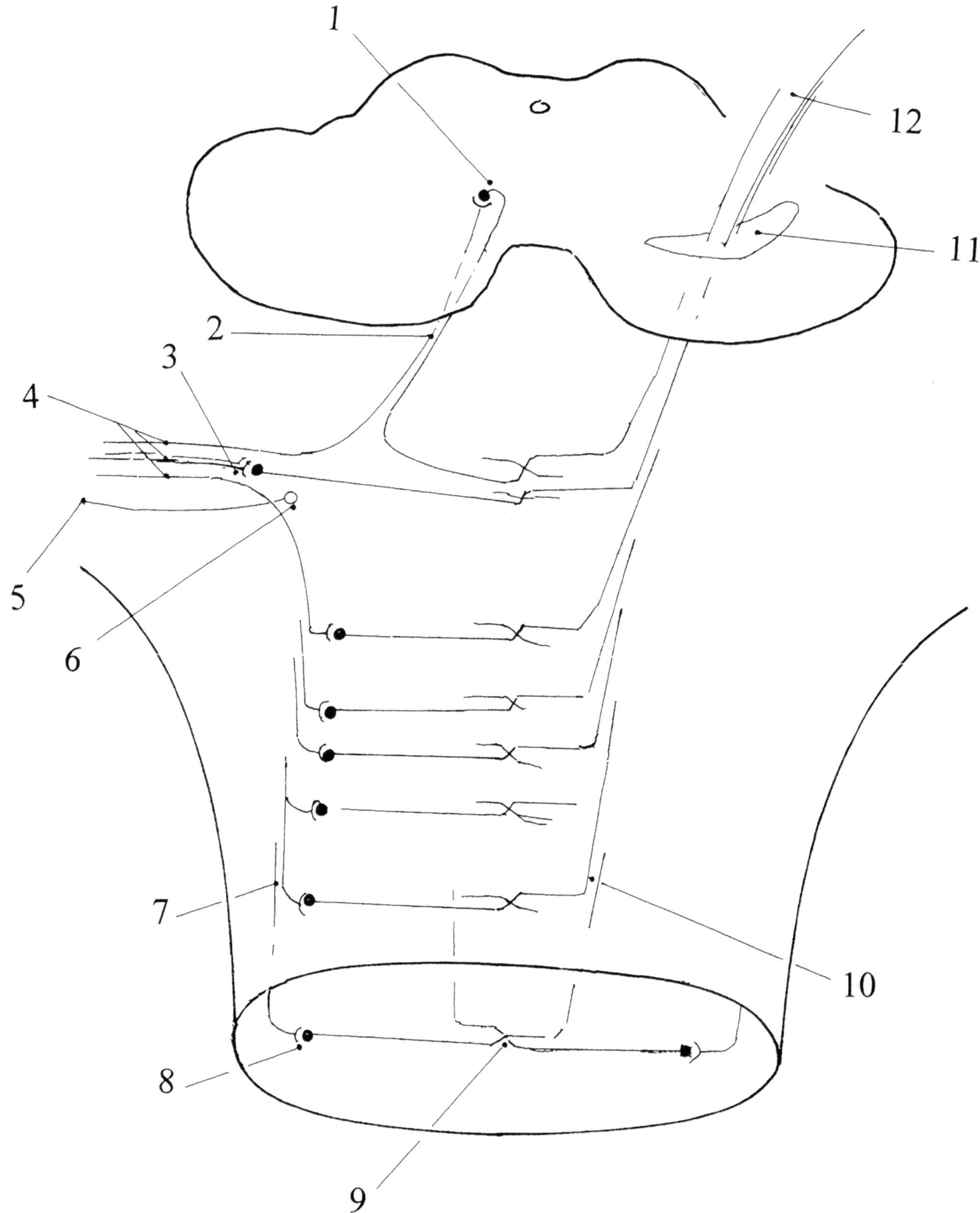

Fig. 8.10 N. trigeminus and its lemniscus medialis segment. (*1*) Nucleus tractus mesencephalicus nervi trigemini, (*2*) tractus mesencephalicus nervi trigemini, (*3*) nucleus sensorius principalis nervi trigemini, (*4*) efferent fibers of ganglion trigeminale, (*5*) radix motoria nervi trigemini, (*6*) nucleus motorius nervi trigemini, (*7*) tractus spinalis nervi trigemini, (*8*) nucleus tractus spinalis nervi trigemini, (*9*) decussatio lemniscorum, (*10*) and (*11*) lemniscus medialis, (*12*) continuation into thalamus

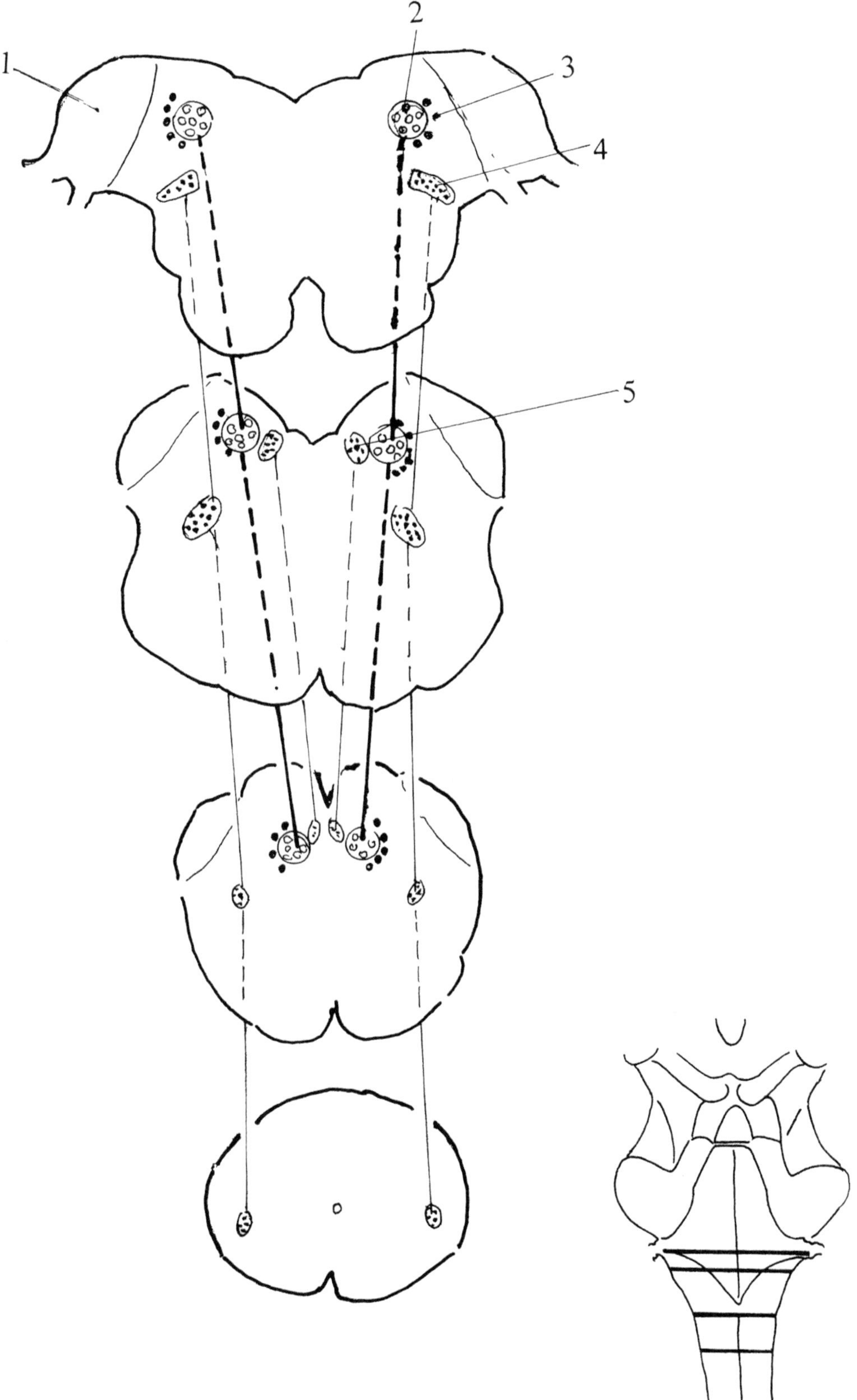

Fig. 8.11 Close topographical relationships of pedunculus cerebellaris inferior with nucleus et tractus solitarius. (*1*) Pedunculus cerebellaris inferior, (*2*) tractus solitarius, (*3*) nucleus tractus solitarii, (*4*) nucleus ambiguus, (*5*) nucleus dorsalis nervi vagi

Résumé of Figs. 8.8, 8.9, and 8.10: Lemniscus medialis contains fiber continuations of fasciculus gracilis et cuneatus and of n. trigeminus. Here are no irregular connections or roundabout courses, in contrast to tractus spinothalamicus. Irregular connections may be the reason for the only transitory analgetic effect of surgical chordotomy (this effect is only in a short duration, neurosurgical experience) of tractus spinothalamicus

9 Cranial Nerves

9.1 Principles of the Embryonic Developments: Survey (Fig. 9.1)

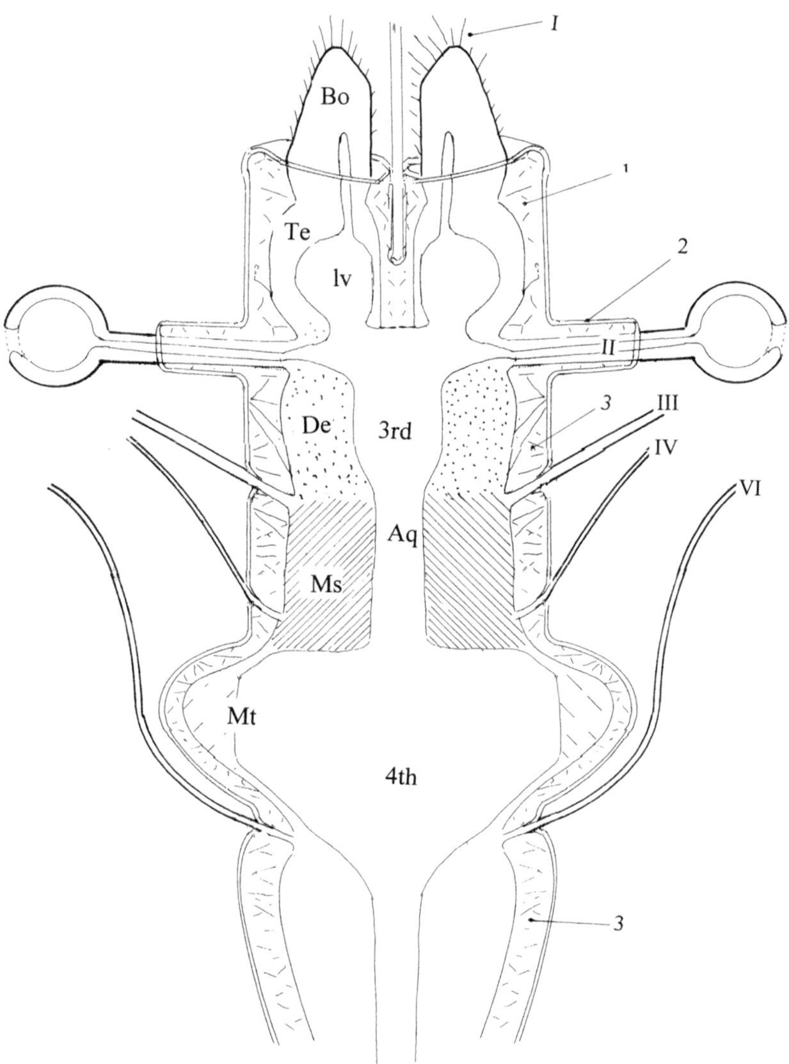

Fig. 9.1 Fila olfactoria presents n. olfactorius. Bulbus and tractus olfactorius are cerebral segments of telencephalon (rhinencephalon) in all craniata. The so-called n. opticus is a segment of diencephalon in all craniata. Bulbus et tractus olfactorius and "n." opticus (II) penetrates its arachnoid cisterns and is bulging into the subdural space. *Bo* bulbus olfactorius, *Te* telencephalon, *De* diencephalon, *Ms* mesencephalon, *Mt* metencephalon, *lv* lateral ventricle. 3rd, Aq and 4th are further ventricular compartments; (*1*) cisterna olfactoria, (*2*) arachnoid membrane of cisterna optica, (*3*) cisternae basales

W. Seeger, *Evolution of the Central Nervous System of Craniata and Homo*, https://doi.org/10.1007/978-3-030-15216-1_9

9.2 Cranial Nerves: Details (Figs. 9.2, 9.3, 9.4, 9.5, 9.6, 9.7, 9.8, 9.9, 9.10, 9.11, and 9.12)

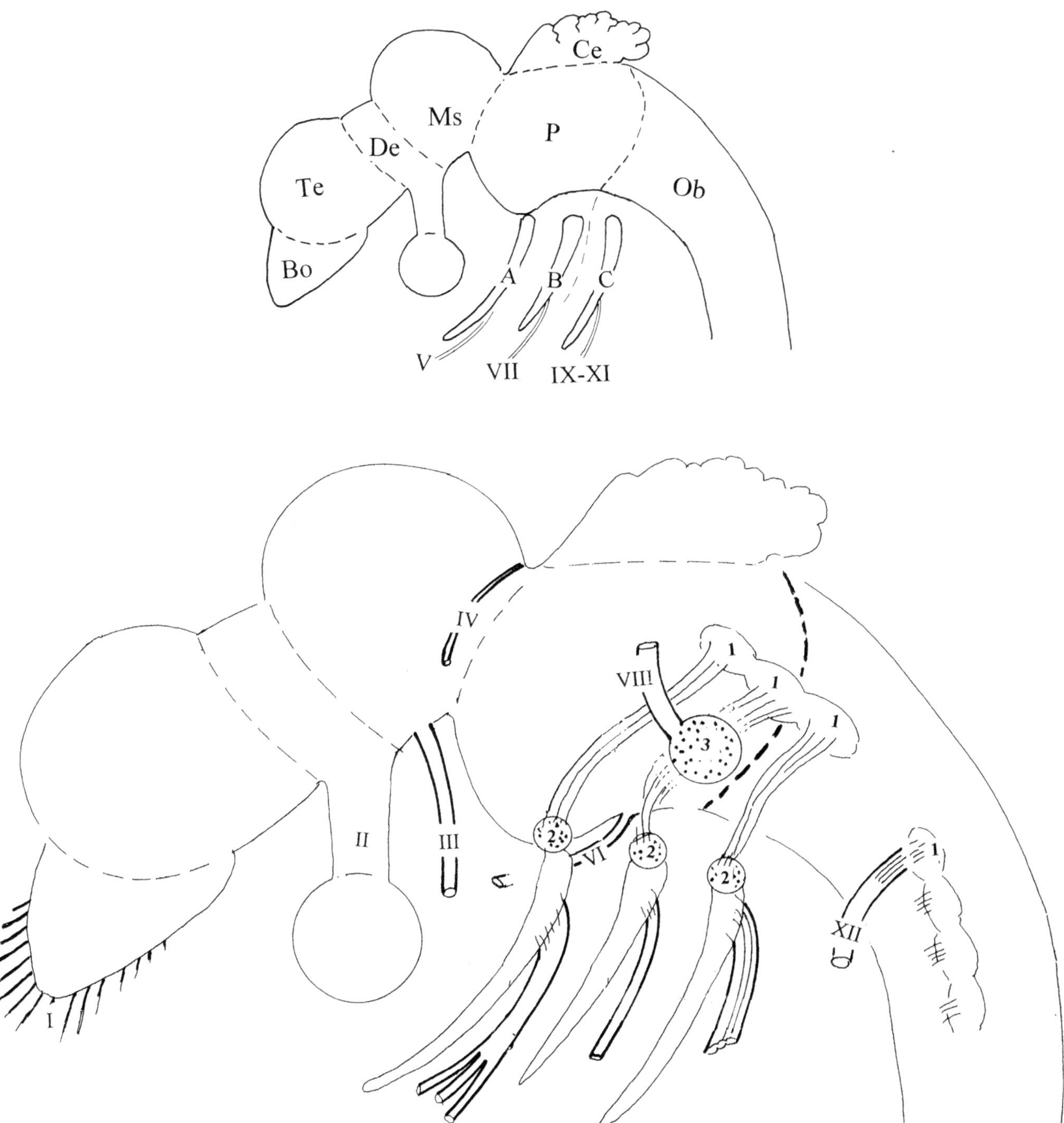

Fig. 9.2 *Caudal (and rostral) cranial nerves of terrestric craniata. Principles of the embryonic transformation of neural crests, ectodermal plates, and nerves of branchial arches into cranial nerves.* (A): first branchial arch (n. V), (B): second branchial arch (n. VII), (C): third branchial arch (nn. IX–XI). *Bo* bulbus olfactorius, *Te* telencephalon, *Di* diencephalon, *Ms* mesencephalon, *Ce* cerebellum, *P* pons, *Ob* medulla oblongata. Rostral cranial nerves: I–IV and VI. Caudal cranial nerves: V n. trigeminus, VII n. facialis, VIII n. statoacusticus, IX–XI n. glossopharyngeus, vagus et accessorius, XII n. hypoglossus. (*1*) Neural crests, (*2*) ectodermal plates connected with (*1*), (*3*) acoustic ectodermal plate connected with neural crests. Distal from (*2*): branchial arches A–C and cranial nerves V, VII, and IX–XI

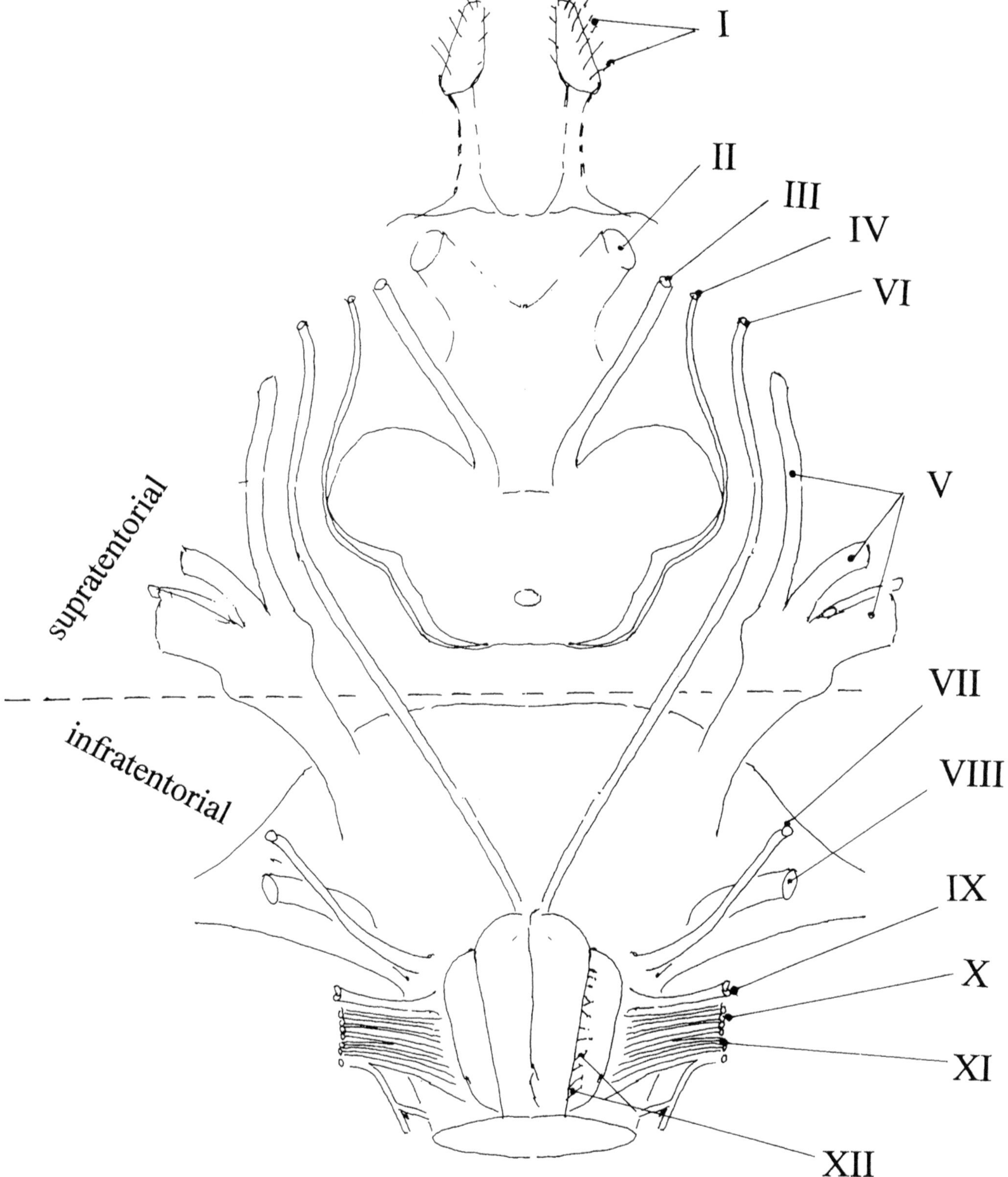

Fig. 9.3 *Rostral and caudal cranial nerves of terrestric craniata. Example homo.* Synopsis

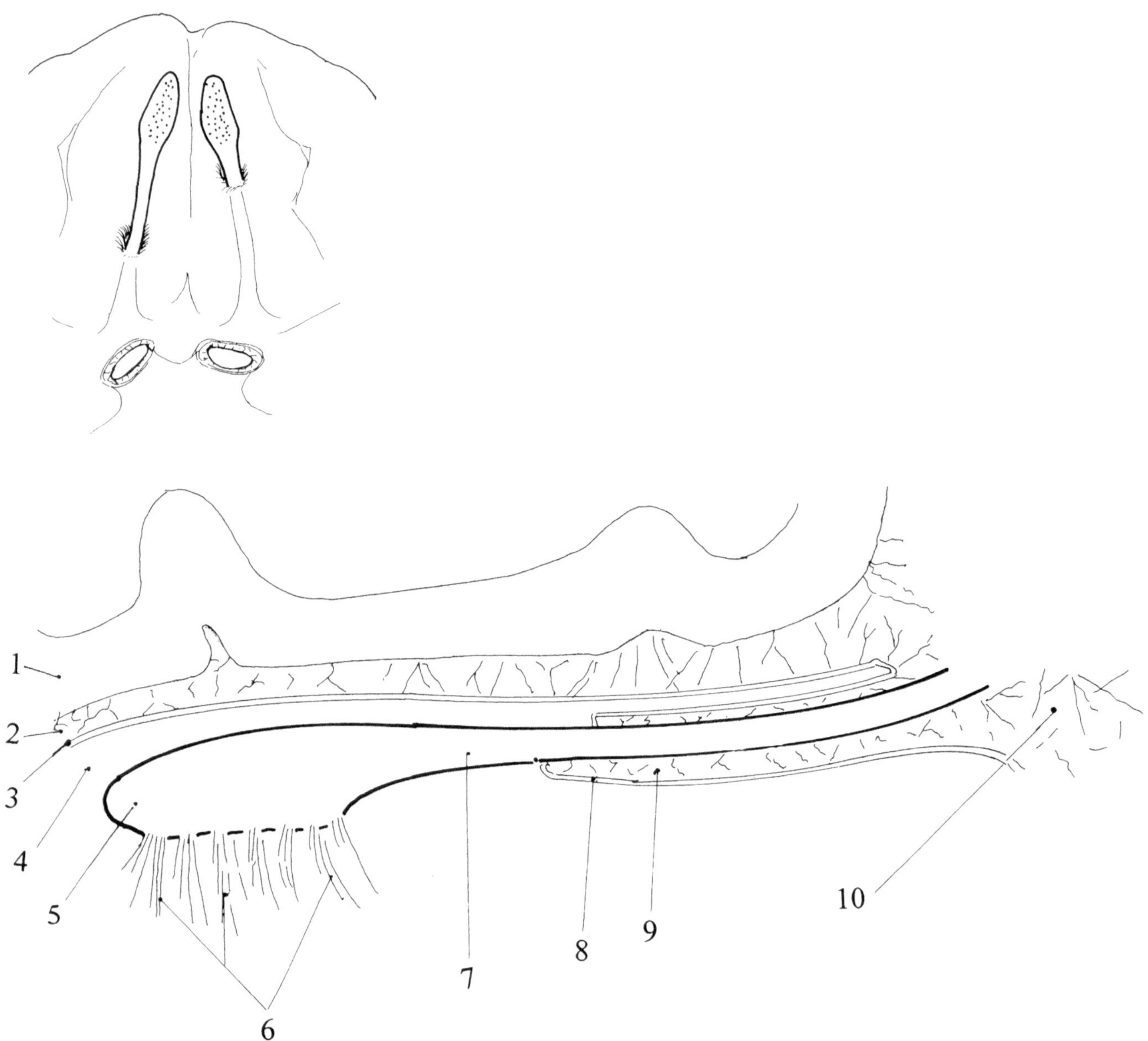

Fig. 9.4 *Tractus et bulbus olfactorius and fila olfactoria.* (*1*) Cortex, (*2*) subarachnoid space, (*3*) outer arachnoid wall, (*4*) subdural space, (*5*) bulbus olfactorius, (*6*) fila olfactoria, (*7*) tractus olfactorius, (*8*) arachnoid wall of cisterna olfactoria, (*9*) cisterna olfactoria, (*10*) cisterna optica

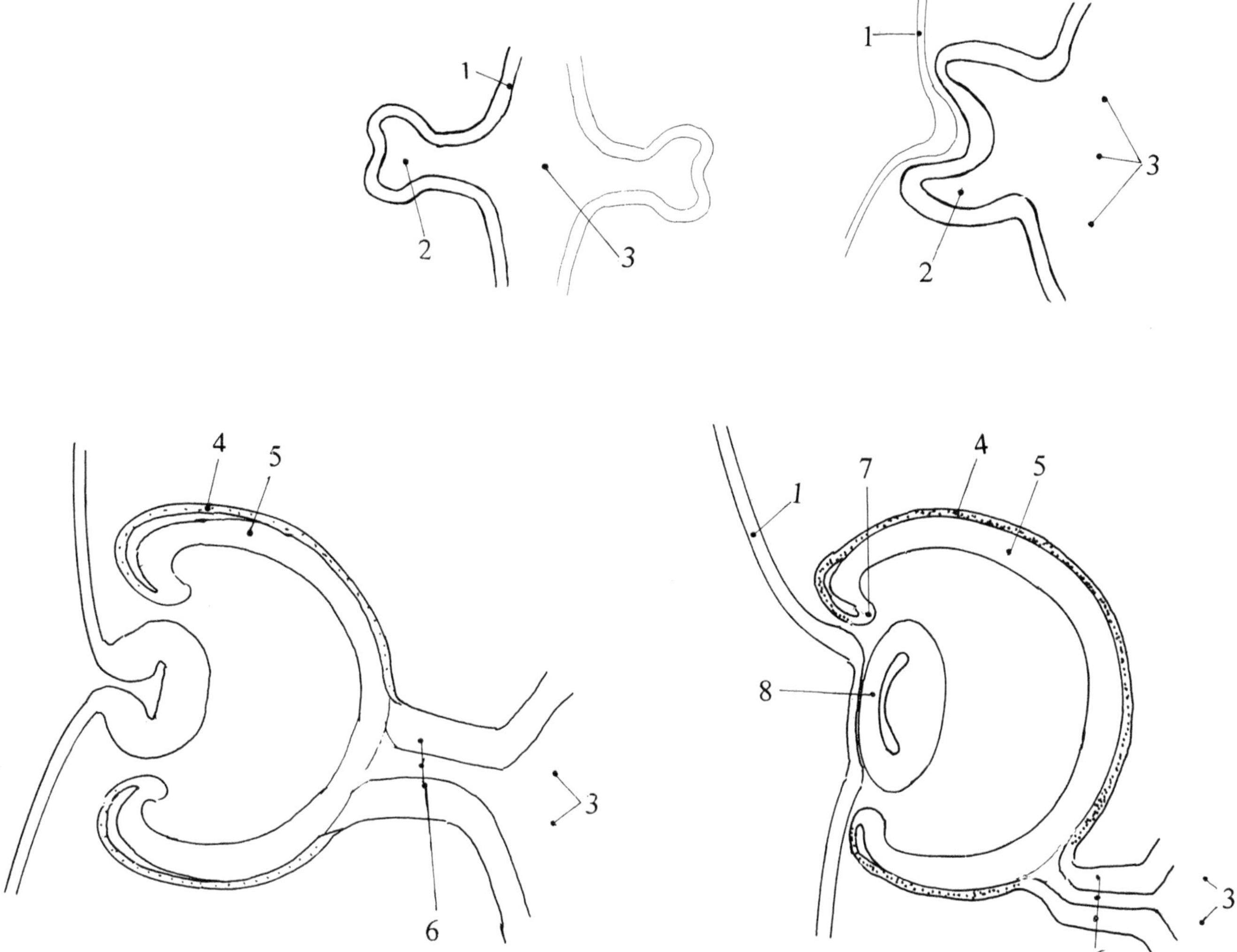

Fig. 9.5 *Fasciculus ("n.") opticus. Embryogenesis*. Fasciculus opticus is a subdural-extradural evacuation of diencephalon, together with retina. (*1*) Ectoderm, (*2*) optic vesicle, (*3*) third ventricle, (*4*) pigment layer, (*5*) retina, (*6*) fasciculus opticus, (*7*) iris, (*8*) lens

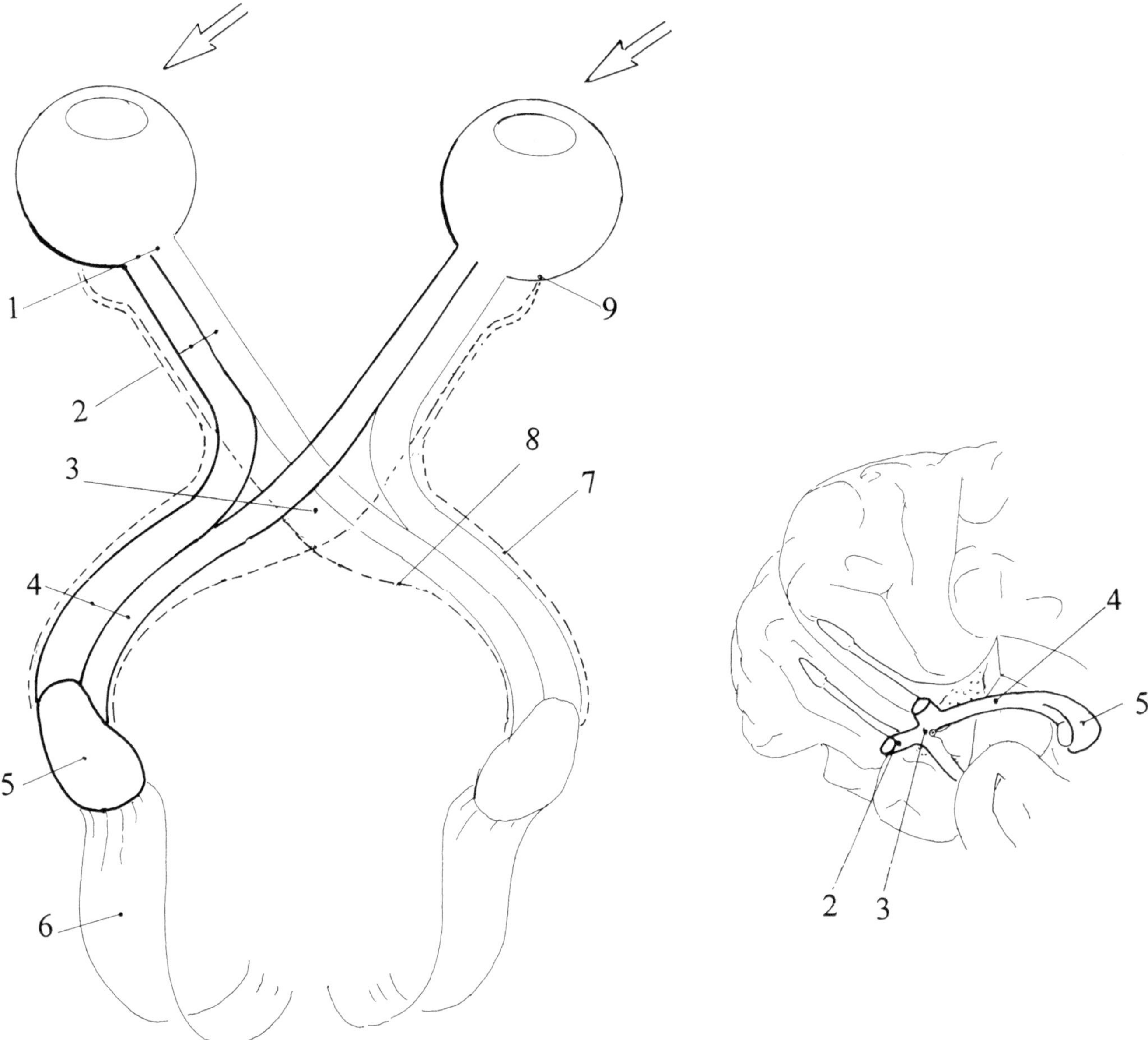

Fig. 9.6 *Optic fiber system of hominoides and homo*. Light beam from one side (arrows) is projected into the contralateral segment of both retinae and into the contralateral lobus occipitalis. Central visual functions (9) are located bilateral (dotted). (*1*) Discus nervi optici, (*2*) optic stalk, (*3*) chiasma opticum, (*4*) tractus opticus, (*5*) corpus geniculatum laterale, (*6*) radiatio optica (for area 17 of cortex), (*7*) fibers from macula lutea, (*8*) as (*7*), (*9*) macula lutea

Fig. 9.7 *Continuation. Motor nerves of bulbi oculi (and iris)*, nn. III, IV, and VI. (*1*) m. rectus medialis, (*2*) m. rectus inferior, (*3*) m. rectus superior, (*4*) m. obliquus inferior, (*5*) m. levator palpebrae superioris, (*6*) m. obliquus superior, (*7*) m. rectus lateralis, *M* mesencephalon, *Ob* pontomedullary rim (isthmus rhombencephali). Constrictor pupillae: N. III, parasympatic component, dilatator pupillae: ganglion ciliare (sympathicus)

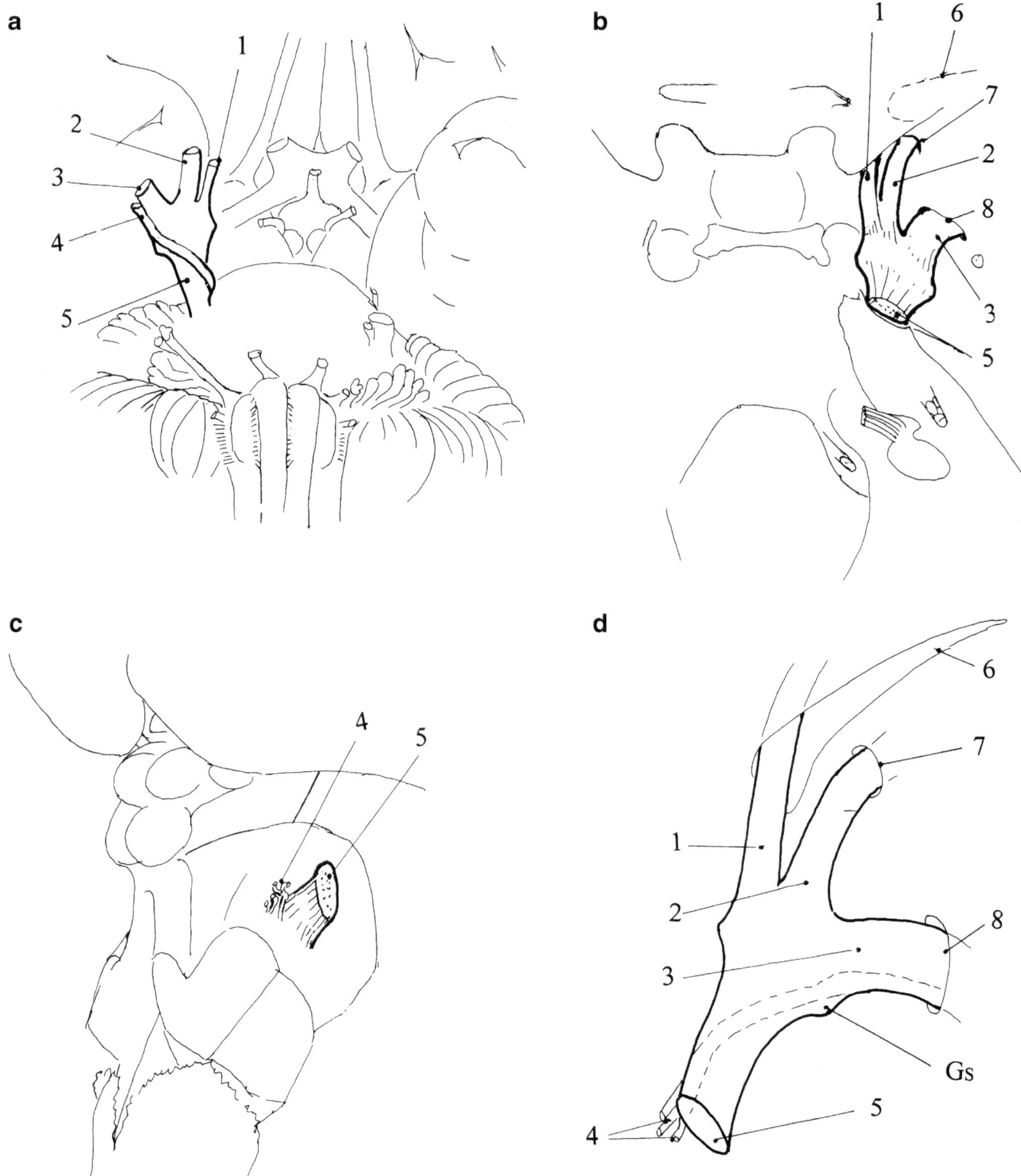

Fig. 9.8 *N. trigeminus, n. V.* (**a**) Basal viewing direction, (**b**) cranial base and bony penetration points of n. trigeminus, (**c**) dorsolateral viewing direction, (**d**) sectional magnification of b, (*1*) n. ophthalmicus, (*2*) n. maxillaris, (*3*) n. mandibularis, (*4*) ramus motoricus, often mixed with sensoric fibers (Rhoton, personal communication to the author in Zürich), (*5*) main trunk, (*6*) fissura orbitalis superior, (*7*) canalis rotundus, (*8*) foramen ovale, Gs ganglion semilunare (Gasseri). (Motor nerves: Radix motoria, n. tensor tympani)

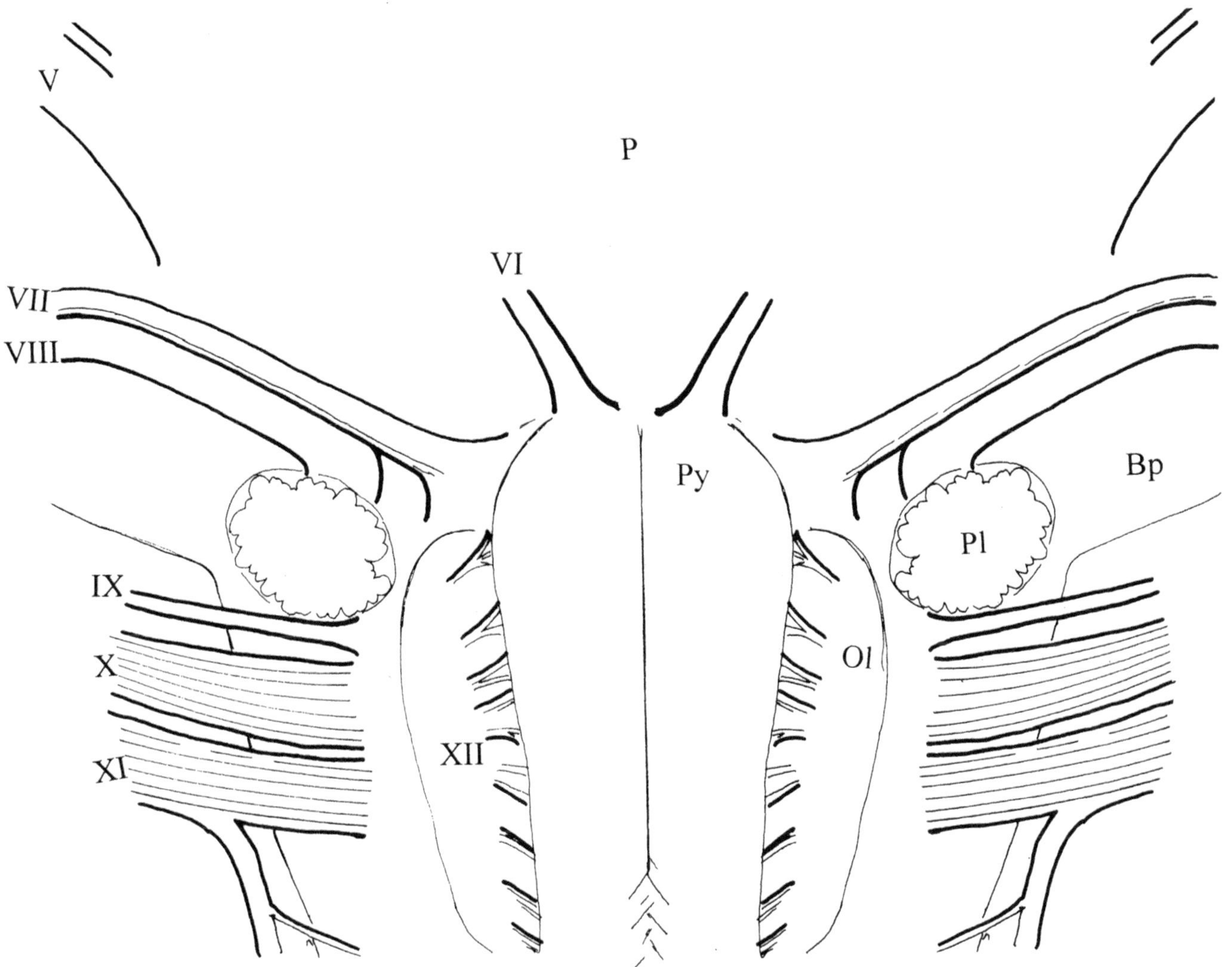

Fig. 9.9 *Caudal cerebral exit region of cranial nerves.* V, N. trigeminus, VII n. facialis, n. stapedius and intermedius, VIII n. statoacusticus, IX–XI nn. glossopharyngeus, vagus et n. accessorius, XII N. hypoglossus, *P* pons, *Bp* brachium pontis (pedunculus cerebellaris medius), *Py* tractus pyramidalis, *Ol* oliva, *Pl* plexus chorioideus

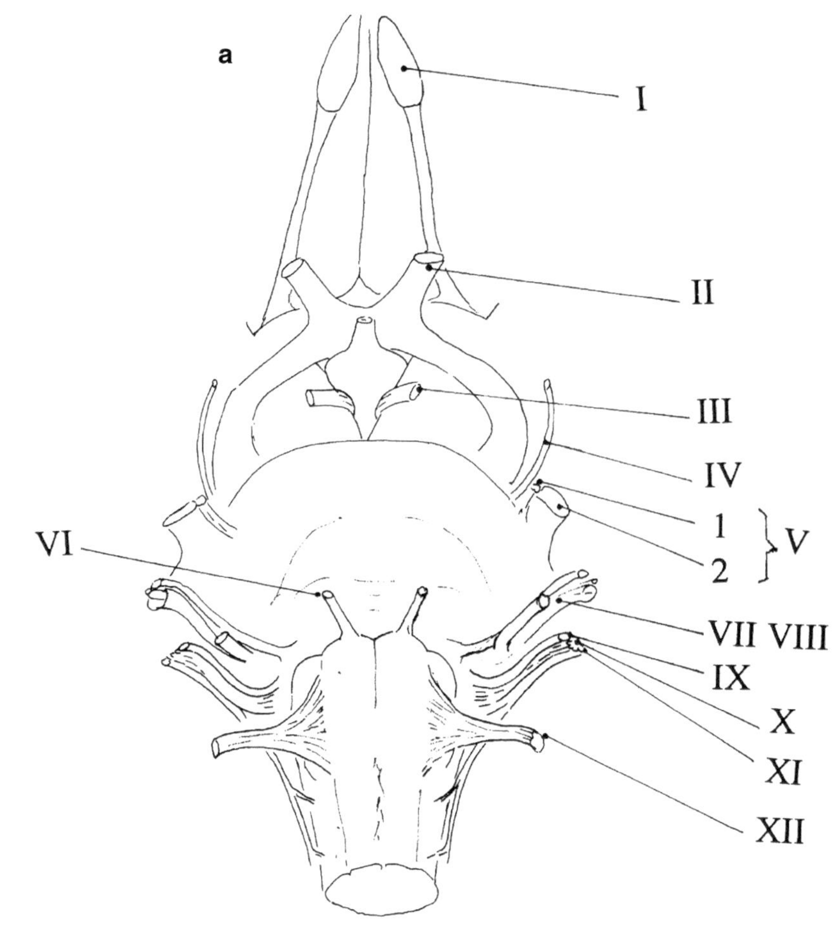

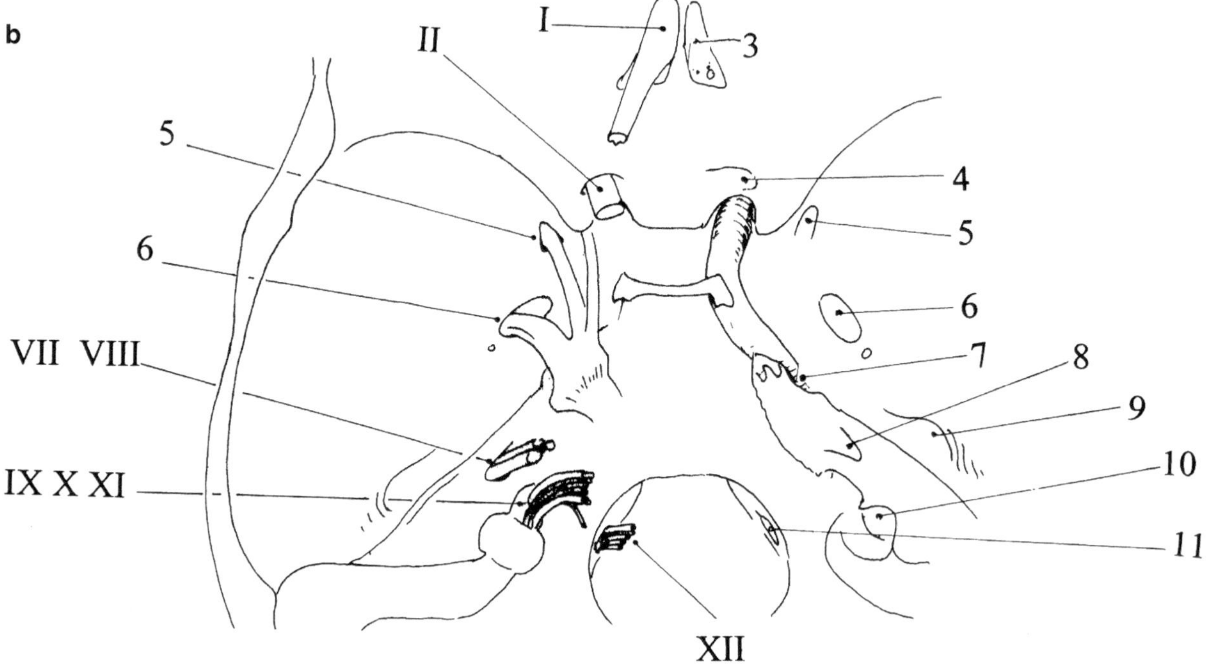

Fig. 9.10 *Cranial nerves and basis cranii interna*. (**a**) Cerebral exit points of cranial nerves (**b**) bony exit points at basis cranii interna. (*1*) Radix motoria N. V, (*2*) radix sensoria N. V, (*3*) lamina cribrosa, (*4*) foramen opticum, (*5*) canalis rotundus enclosing N. maxillaris, (*6*) foramen ovale enclosing N. mandibularis, (*7*) incisura trigeminalis, (*8*) porus acusticus internus, (*9*) eminentia arcuata enclosing labyrinth, (*10*) foramen jugulare, (*11*) canalis hypoglossi

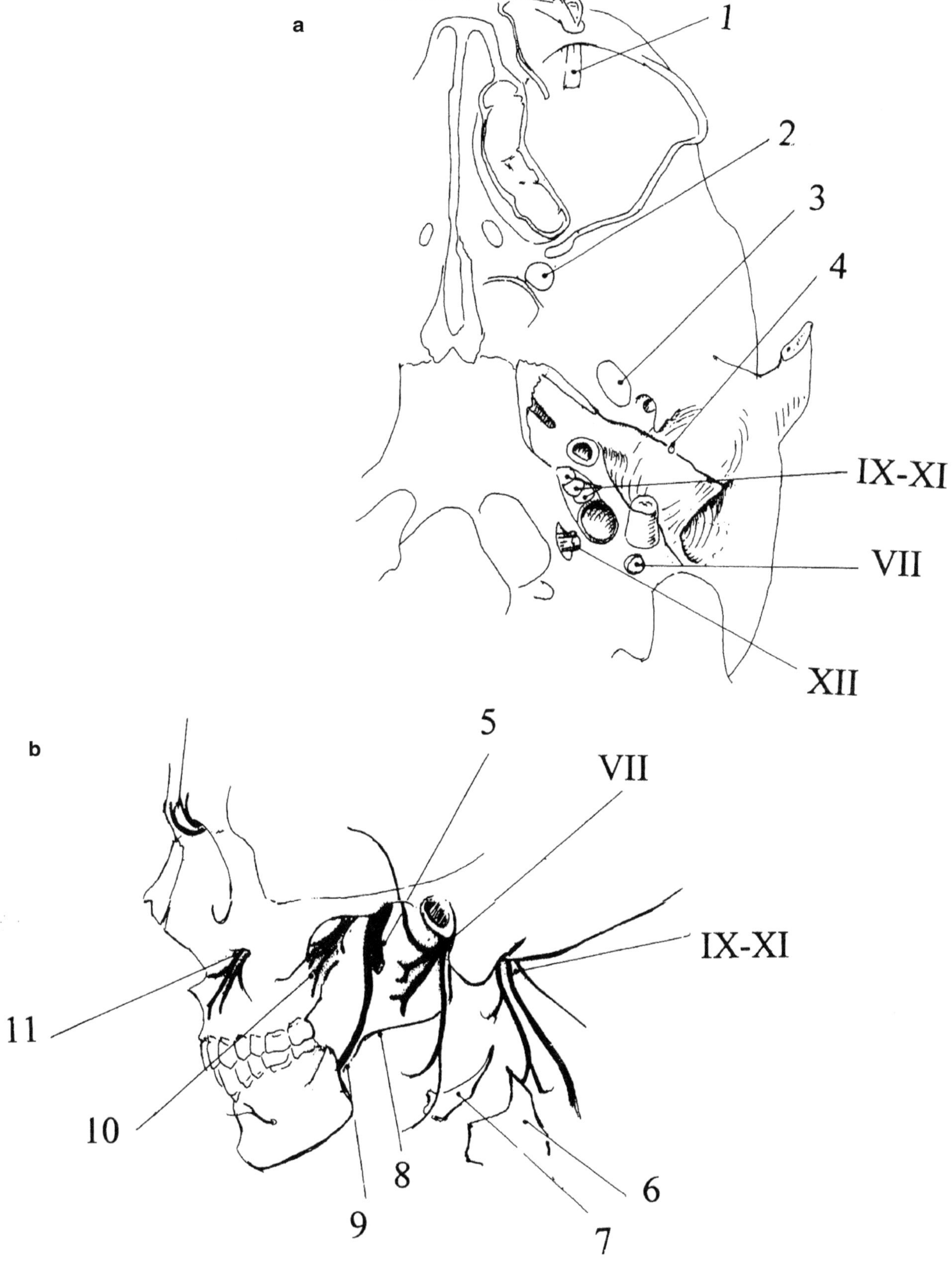

Fig. 9.11 *Cranial nerves and basis cranii externa*. (**a**) Basis cranii externa, penetration of cranial nerves. (**b**) Lateral viewing direction. (*1*) N. supraorbitalis, (*2*) canalis rotundus/N. maxillaris, (*3*) foramen ovale/N. mandibularis, (*4*) fissura petrotympanica/chorda tympani, (*5*) N. mandibularis, (*6*) cartilago thyreoidea, (*7*) os hyoideum, (*8*) chorda tympani distal from (*4*), (*9*) N. lingualis, (*10*) branches of N. maxillaris, (*11*) branches of N. infraorbitalis

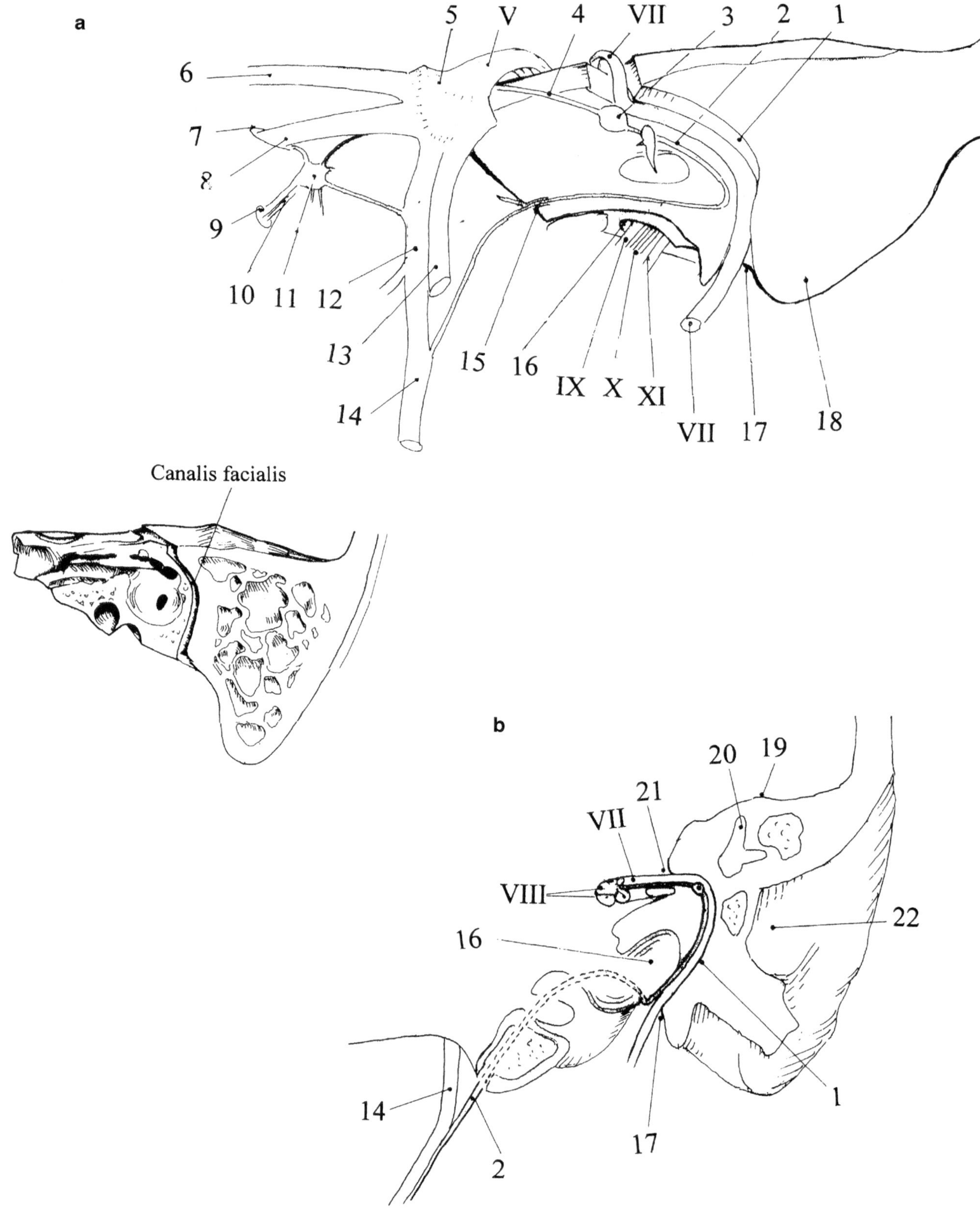

Fig. 9.12 *Caudal cranial nerves at special craniobasal structures*. (**a**) Anterior from canalis facialis, (**b**) posterior from canalis facialis. (*1*) Canalis facialis, (*2*) chorda tympani, (*3*) ganglion geniculi, (*4*) n. intermedius, (*5*) ganglion semilunare (Gasseri), (*6*) n. ophthalmicus, (*7*) foramen rotundum, (*8*) n. maxillaris, (*9*) canalis pterygoideus, (10) n. pterygoideus, (*11*) ganglion sphenopalatinum, (*12*) n. mandibularis, (*13*) radix motoria nervi trigemini, (*14*) n. lingualis, (*15*) fissura petrotympanica and chorda tympani, (*16*) foramen jugulare, (*17*) foramen stylomastoideum, (*18*) processus mastoideus, (*19*) eminentia arcuata, (*20*) ductus semilunaris superior, (*21*) porus acusticus internus, (*22*) meatus acusticus externus

Part III

Morphologic Maturity and Immaturity

Pisces, Amphibia, Reptilia

Archaic fossil sharks present a similar size of neurocranium as recent archaic sharks. Similar findings present recent and fossil amphibians and reptiles. The archaic immaturity and maturity of cerebrum are not to prove. Recent encephalitis presents well-matured systematic fiber structures with avoiding roundabout long fiber systems of medulla spinalis, myelencephalon, mesencephalon, and diencephalon of all craniata, including homo. Even small lesions are not tolerated. Small well-defined location of surgical manipulations at basal ganglia can restore functional defects, e.g., at Parkinson's disease of homo. The size of cavum cranii of the fossil saurian reptiles had more than hundred million years of time to differentiate its brain systematic. This may explain that the size of cavum cranii of saurian reptiles was smaller than that of other reptiles, avoiding roundabout and duplicated fiber systems.

Aves

Allocortex of telencephalon is highly magnified in comparison with reptiles (see testudo Fig. 4.1 and goose Fig. 3.4). Allocortex of all craniata presents high variable cellular structures, depending on localizations and functions.[38] The magnified cerebellum of aves is partly well organized. Its development arrived independent of the even magnified cerebellum of mammalia.

Mammalia and Homo

Instead of magnification of allocortex according to aves, mammalia created a new kind of cortex of telencephalon: **Neocortex**. Its tissue presents seven variable layers, depending on localizations and functions. Granule cell layers and pyramidal cell layers are predominant in contrast to allocortical structures. The high progredient magnification of neocortex is combined with a stepwise diminution of allocortex of mammalia. Without allocortex, homo would lose consciousness, as well as after large defects of formatio reticularis of mesencephalon and upper pons, which are connected with allocortex (tractus mammillothalamicus and amygdala–basal ganglia). Magnification of neocortex is combined with a progredient development of gyrification. The highest grade of neocortical gyrification presents Cetacea, especially dolphins, more than other mammalia and homo (Fig. 3.9). Makroosmate mammalia (canis, equus—Fig. 3.7) present a well-preserved rhinencephalon combined with high-grade gyrification of neocortex. Microosmates presents hominoides, homo, and most cetaceans. Small lesions of allocortex (fornix, corpus mamillare, stria longitudinalis) may endanger all psychological functions in homo. Large unilateral lesions of the frontal neocortex at the nondominant hemisphere are usually tolerated without psychological deficits, if its allo- and mesocortex were preserved. Special psychological functions (speech, reading, writing, calculating) are located in special regions of the dominant hemisphere,[39] especially parietal and parietotemporal regions. It doesn't tolerate lesions well. Mathematical and other abstract thinking are localized bifrontal. They are not endangered by unilateral frontal lesions, if allocortical and

[38] Stephan (1975)

[39] Motor speech area(s) may be located frontal in the dominant or nondominant hemisphere

mesocortical (striae and cingular) structures are preserved.[40] Extirpation of temporal allo- and mesocortex of the anterior and middle segments (amygdalohippocampectomy) is tolerated without psychological deficits, if the contralateral parts are intact. Visual and acoustic centers are located in both hemispheres.

Even in homo, most associative fibers are archaic U-fibers with multiple less systematic long connections and only roundabout courses. Long associative fascicles with avoiding of roundabout fibers at centrum semiovale are small. Not well proved are the mammalian relationships of the archaic U-fibers with roundabout fibers and small associative fascicles of centrum semiovale. The well-developed gyrification of telencephalon of cetacea is more developed than in hominoides and homo. But since the early terrestrial mammalia have adapted to submarine conditions, cetaceans were living 40–60 million years. Homo were living only less than 3 million years, changing from sylvestric lives to the more problematic lives of homo erectus. During these short times the rapid increase of telencephalon and reduction of allocortex of homo happened. **Telencephalon of recent homo is an early morphologic stage of the cerebral development**. Diencephalon, mesencepalon, and rhombencephalon are common archaic similar structures of all craniata, including homo. Roundabout and duplicated fiber connections here are avoided or reduced. This systematic architecture is combined with a high vulnerability of its structures, in contrast to the frontal neocortex of the nondominant hemisphere of homo. A high vulnerability presents projection systems except the archaic corpus callosum. Highly vulnerable are the special psychological centers of the parietal lobe of the dominant hemisphere. This region is smaller than the nondominant parietal region (Fig. 9.11). It may indicate that this area is more systematically structured than the contralateral nondominant parietal lobe.

[40] Experiences in neurosurgery, psychologically tested by B.-O. Hütter. Personal communication to the author in Aachen 1999 and 2000

10 Recent Neocortex

10.1 Neocortical Areas (Figs. 10.1, 10.2, and 10.3)

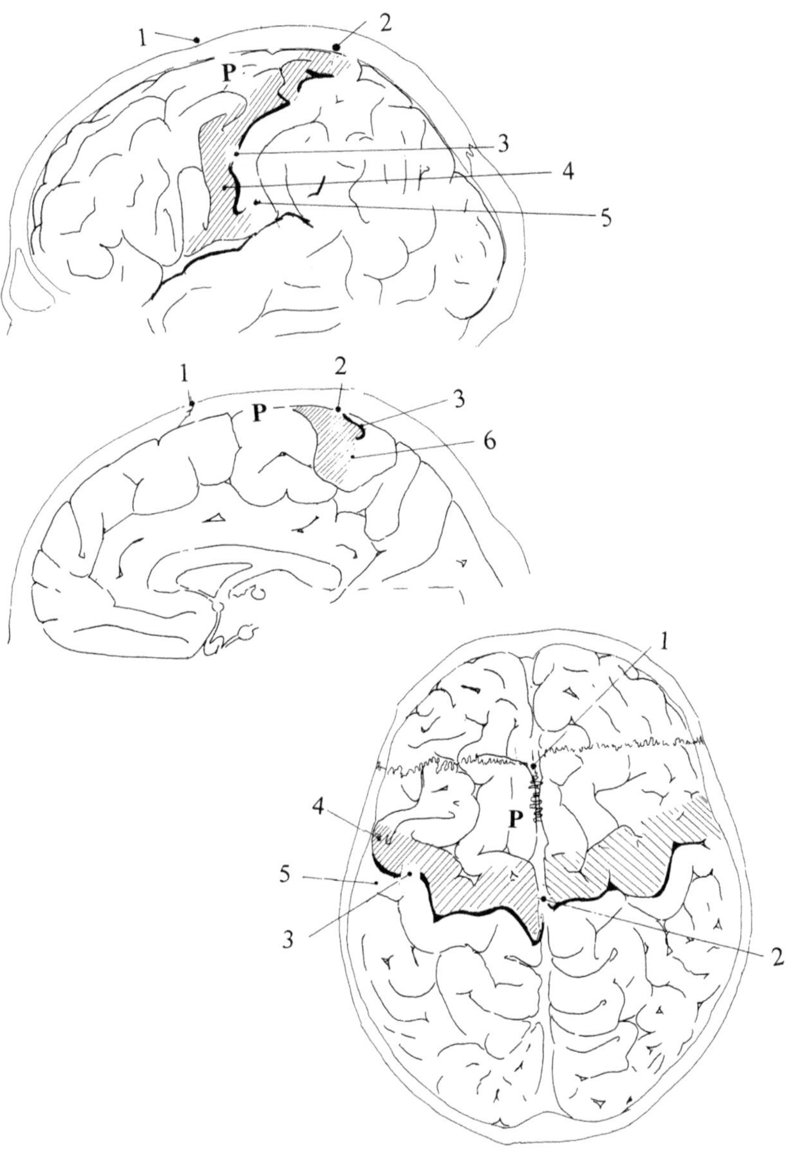

Fig. 10.1 *Cortical motor regions. P* premotor planning area, (*1*) bregma, (*2*) highest point of sulcus centralis, (*3*) sulcus centralis, (*4*) gyrus praecentralis, (*5*) gyrus postcentralis, (*6*) gyrus paracentralis

W. Seeger, *Evolution of the Central Nervous System of Craniata and Homo*, https://doi.org/10.1007/978-3-030-15216-1_10

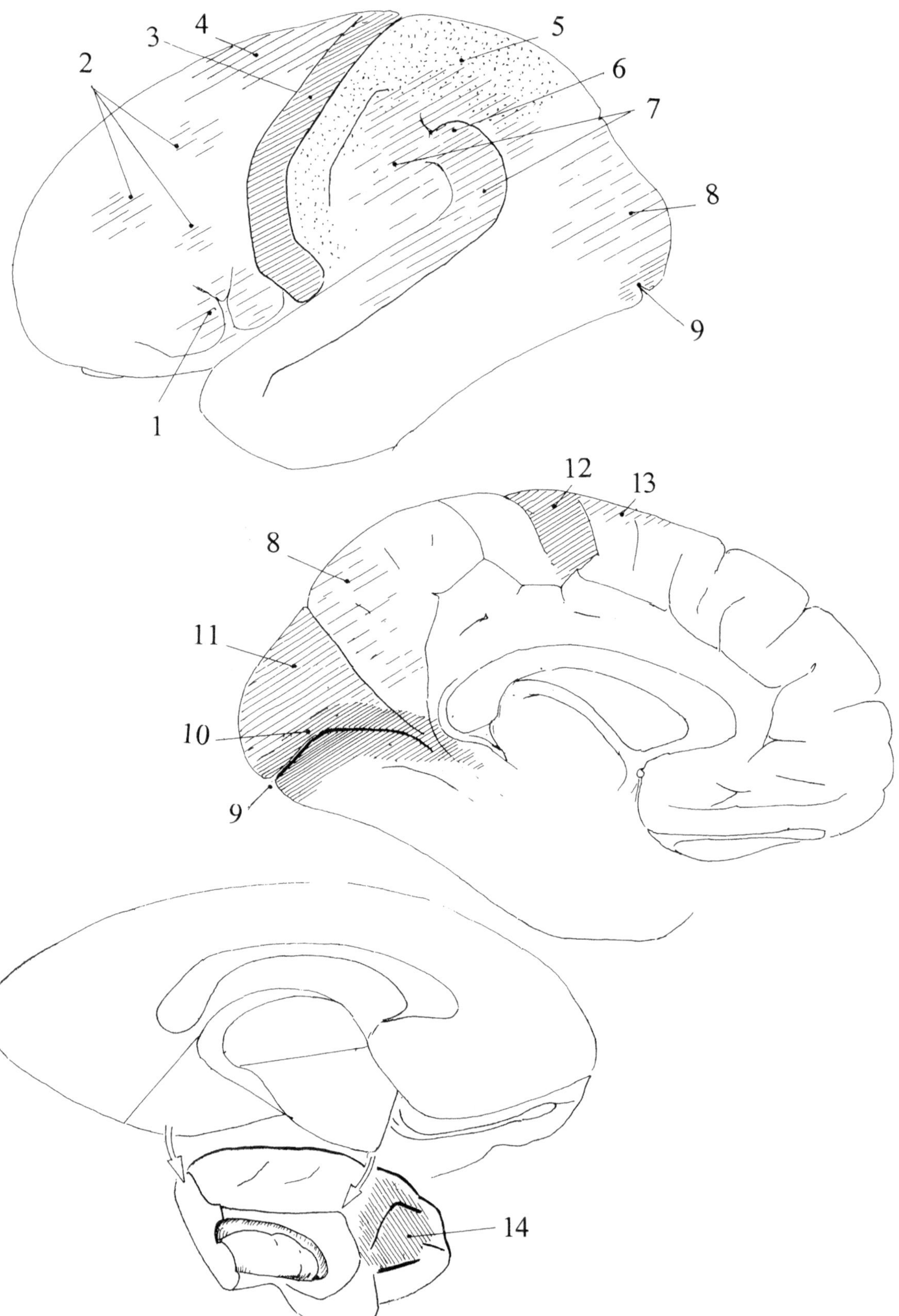

Fig. 10.2 *Neocortical centers for special psychological and functions of the dominant hemisphere and motor/sensoric centers of both hemispheres*. Exact individual defining by performing neurophysiological methods. (*1*) Broca's area, (*2*) variable uni- or contralateral motor speech areas, (*3*) gyrus precentralis (motor area), (*4*) premotor planning area, (*5*) sensoric area, (*6*) gyrus angularis (Wernicke), (*7*) centers for speech memory and speech sensory, reading, writing, calculating, and control of the contralateral hand, (*8*) secondary optic area, (*9*) sulcus calcarinus (posterior end), (*10*) central optic area (sulcus calcarinus), (*11*) primary optic area, (*12*) motor center for both legs and for bladder, (*13*) sensory center, (*14*) temporal gyri transversi, acoustic center (Heschl)

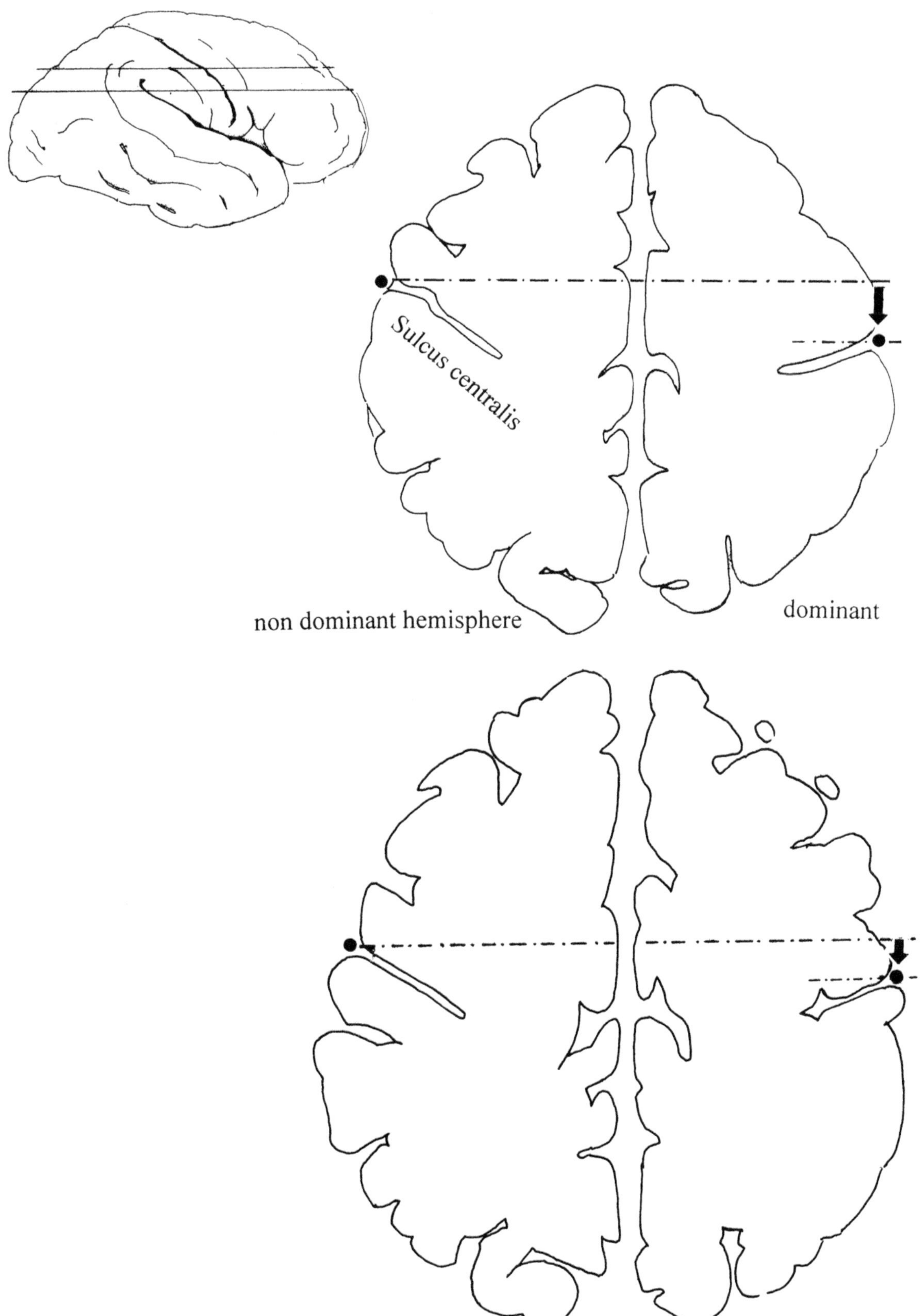

Fig. 10.3 *Usual asymmetry of hemispheres. MRT. At the dominant hemisphere sulcus centralis is retropositioned. Parietal lobe is smaller than at the nondominant hemisphere.* **This diminution of neocortical areas of elementary psychological functions may be a reference for the beginning of a morphological development of systematic morphologic structures with reduction of roundabout and unnecessary fiber structures and cells**

10.2 Paleoanthropological Aspects (Figs. 10.4 and 10.5)

Archeological and modern imaging methods are the base for defining cavum cranii. Size of cavum cranii is nearly identical with brain volume. Exact measures are hindered by the incomplete preservation of fossil skulls and by the high variability of the human brain. von Bischoff [16] proved the recent brain weights of 559 adult male with 1000 g to maximal 1925 g (middle weight 1362 g) and 347 adult females with 820 g to maximal 1565 g (middle weight 1219 g). **The brain weights of *Homo heidelbergensis*, homo of Steinheim, and *Homo neanderthalensis* differ not significant from recent homo** (see Fig. 10.2). **But 2–3 million years before, the brain weights of archaic homines were more and more enlarged onto the development of *Homo heidelbergensis*. This recent type of skull volume is existing since 350,000 years. But nothing is known about the immature or mature of neocortical structures. A reference to beginning maturity of the dominant parietal cortex may be possible, if the right-left difference of cavum cranii in archaic skulls would be proved using modern images** (see Figs. 10.2, 10.3, 10.4, 10.5, and 10.6)**.**

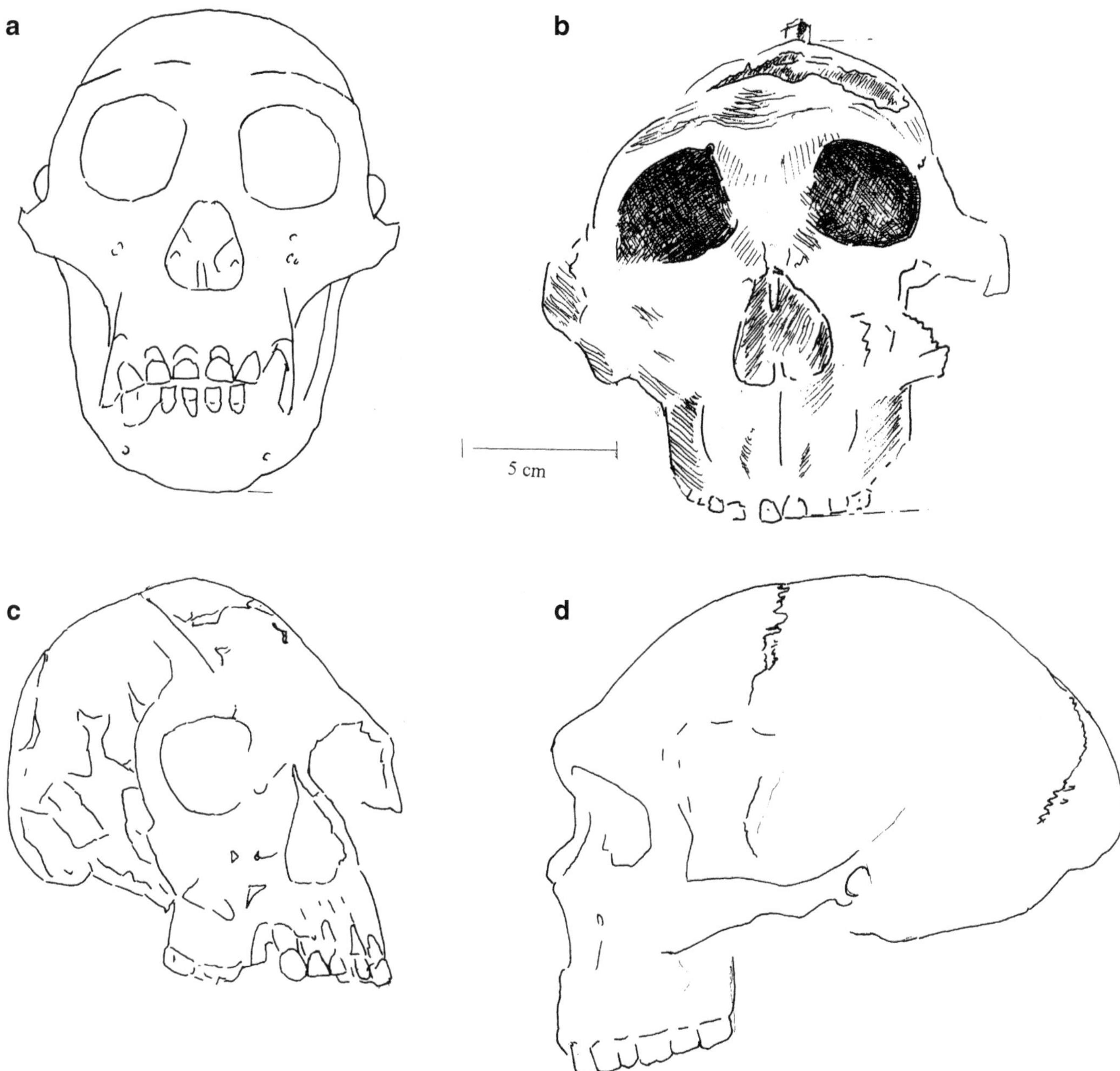

Fig. 10.4 (**a**) *Cranium of recent pan troglodydes (chimpanzee)* for comparison with archaic human skulls. Pan (Pan: Greek and roman god of forest and herdsmen) is the next relative of homo. (**b**) *Australopithecus boisei* (Tattersall and Delson [14], p. 5. Indian ink copy of the author). 2.3–1.6 million years. (**c**) *Homo habilis* (Johanson and Blake [2], p. 86), 2 million years, brain volume 510 ccm. (**d**) *Homo heidelbergensis*, 350,000 years, brain volume 1220 ccm. After *Homo heidelbergensis* the rapid extension of the human brain was stopped

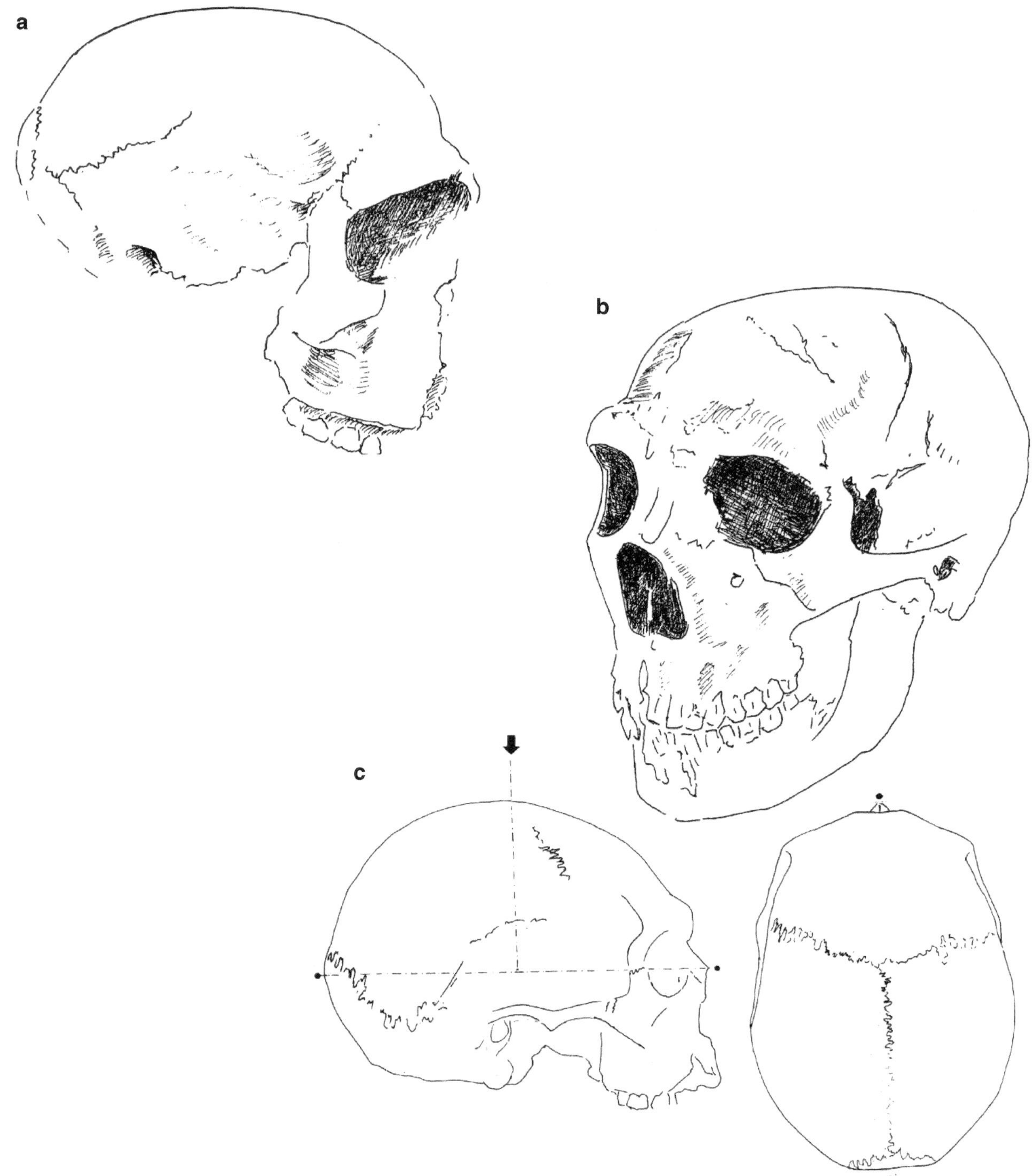

Fig. 10.5 *Cranium of archaic Homo sapiens.* (**a**) Cranium of Steinheim, Germany, 250,000 years (Tattersall and Delson [14], p. 9. Indian ink copy of the author), volume of cavum cranii ca. 1100 ccm. (**b**) Cranium of Neanderthal from the Near East: Amud, Israel, 45,000 years (Johanson and Blake [2], p. 216f; Tattersall and Delson [14], p. 10. Indian ink copy of the author), volume of cavum cranii ca. 1100 ccm. (**c**) Neanderthal type with usual dolichocephaly, volume of cavum cranii ca. 1320 ccm (Collection of Dorpat, according to Rauber-Kopsch [6], p. 310, Indian ink copy of the author, viewing direction drawn in *arrow*)

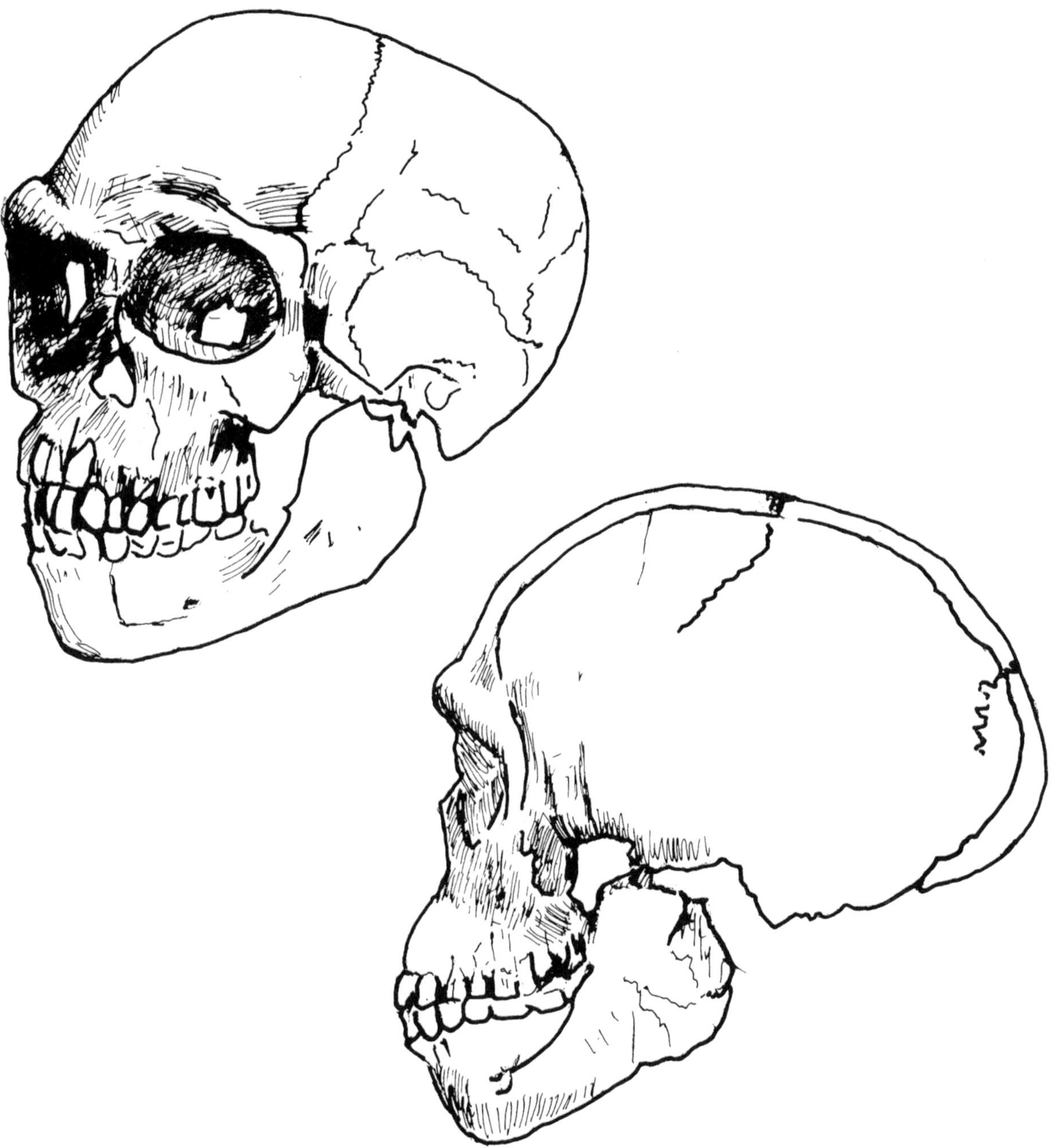

Fig. 10.6 *Addendum*. Modern micro-CTs of 5 fossils of *Homo sapiens* combined, 300,000 years, Jebel Irhoud (Marokko). Oldest fossils of *Homo sapiens* (Hublin et al. [1]. Indian ink copy of the author)

Bibliography

1. Hublin J-J, Ben-Ncer A, et al. Der moderne Mensch ist älter als angenommen. Forschung und Lehre 7/2017 und. 2017. https://doi.org/10.1038/nature22336, MPI für evolutionäre Anthropologie Leipzig.
2. Johanson D, Blake E. Lucy und ihre Kinder. 2. Aufl. München: Elsevier; 2006.
3. Kahle W, Frotscher M. Taschenatlas der Anatomie, 3. Band: Nervensystem und Sinnesorgane. 7. Aufl. Stuttgart: Thieme; 2001.
4. Ludwig E, Klingler J. Atlas cerebri humani. Basel: Karger; 1956.
5. Pilleri G, Giehr M, Kraus C. The structure of the cerebrral cortex of the gangetic dolphin. Z Mikrosk Anat Forsch. 1968;79:373–88.
6. Rauber-Kopsch. Lehrbuch der Anatomie, 7. Aufl., Abt. 2, Knochen, Bänder. Leipzig: Thieme; 1906.
7. Rauber-Kopsch. Lehrbuch der Anatomie, 7. Aufl., Abt. 5, Nervensystem. Leipzig: Thieme; 1907.
8. Rauber/Kopsch. Anatomie des Menschen, Lehrbuch und Atlas, Bd. III. In: Leonhardt H, Töndury G, Zilles K, Hrsg. Nervensystem und Sinnesorgane. Stuttgart: Thieme; 1987.
9. Roth G, Wullimann MF. Evolution der Nervensysteme und der Sinnesorgane. In: Dudel J, Menzel R, Schmidt RF, Hrsg. Neurowissenschaft. Vom Molekül zur Kognition. 2. Aufl. Berlin: Springer; 2001.
10. Seeger W. Atemstörungen bei intrakraniellen Massenverschiebungen. Wien: Springer; 1968.
11. Seeger W. Mikroanatomical aspects for neurosurgeons and neuroradiologists. Wien: Springer; 2000.
12. Spalteholz W. Handatlas der Anatomie des Menschen, Bd. 3, 4. Aufl. Leipzig: Hirzel; 1906.
13. Stephan H. Allocortex, Handbuch der mikroskopischen Anaomie des Menschen, 4. Band. Berlin: Springer; 1975.
14. Tattersall I, Delson E. Ancestors, four million years of humanity. New York: American Museum of Natural History; 1984.
15. von Bischoff TLW. Die Großhirnwindungen des Menschen. München: Verlag kgl. Akademie; 1868.
16. von Bischoff TLW. Das Hirngewicht des Menschen. Bonn: P. Neusser; 1880.

Further Readings

Dudel J, Menzel R, Schmidt RF, Hrsg. Neurowissenschaft. Vom Molekül zur Kognition. 2. Aufl. Berlin: Springer; 2001.

Jefferson TA, Leatherwood S, Webber MA. Marine mammals of the world. Rome: FAO and UNEP; 1993.

Kisia SM. Vertebrates: structures and funkctions. Boca Raton: CRC Press, Taylor & Francis Group; 2017.

Seeger W. Atlas of topographical anatomy of the brain and surrounding structures. Wien: Springer; 1978.

Springer SP, Deutsch G. Linkes – rechtes Gehirn. Funkionelle Asymmetrien. 2. Aufl. Heidelberg: Spektrum-der-Wissenschaft-Verlagsgesellschaft; 1988.

W. Seeger, *Evolution of the Central Nervous System of Craniata and Homo*, https://doi.org/10.1007/978-3-030-15216-1